SCATTERED
MATHERTICLES

Mathematical Reflections

Volume I

Satish C. Bhatnagar

Order this book online at www.trafford.com/08-0273
or email orders@trafford.com

Most Trafford titles are also available at major online book retailers.

Note for Librarians: A cataloguing record for this book is available from Library
and Archives Canada at www.collectionscanada.ca/amicus/index-e.html

Printed in Victoria, BC, Canada.

ISBN: 978-1-4251-7247-3 (sc)
ISBN: 978-1-4251-7248-0 (e)

 www.trafford.com

North America & international
toll-free: 1 888 232 4444 (USA & Canada)
phone: 250 383 6864 ♦ fax: 250 383 6804 ♦ email: info@trafford.com

The United Kingdom & Europe
phone: +44 (0)1865 487 395 ♦ local rate: 0845 230 9601
facsimile: +44 (0)1865 481 507 ♦ email: info.uk@trafford.com

10 9 8 7 6 5 4 3 2 1

FAR MORE THAN A PREFACE

Background

Lately, I have not started on a book or adopted a mathematics textbook without reading author's preface, introduction, or foreword. It aptly applies to my collection of *Reflections*. In the US, I have heard it said that in life, a man must accomplish three things - plant a tree, have a son, and write a book. Perhaps, it defines a simple legacy at the individual level. Nevertheless, it nurtures the environment and human race. Besides two daughters and their children, my 42-year old son has a son and daughter. It makes me feel doubly blessed. Several trees, planted by me, have been flourishing for years now. But, for a number of reasons, a book has evaded me for years.

However, my book part of the saying is a story by itself. It reminds me of my late colleague and good friend, John Abramowich, who used to say that read a book that is 100-year old. But you simply can't rule all the new prints out. There is no litmus test for the longevity of a book, or for that reason, of a human being. All 100-year old books were once one-day old. Nonetheless, Abramowich did influence my thinking. After a certain 'ripe' age, it is wasteful time to keep on reading - creating intellectual constipation. In this age of print pollution, I have come up with a corresponding benchmark on writing - write a book, as if it is going to last for 100 years!

Surprisingly, the time has come to write a preface for my first book. It is a poetic justice in a sense. At age 22, I declined the offer of having my graduate notes on *Solid Geometry* (E. T. Bell) published, while living in India then. My idealistic reason was that it would not be helping the students, as they would run up for the solutions, rather than think and attempt the problems themselves. The event is fresh, though it happened 48 years ago.

Now, I asked these questions from myself: What good these *Mathematical Reflections* are? Who are going to enjoy reading, or benefit from them? What is my claim to their worth? It took me nearly four years to go for publication. These *Reflections* are a good reading for general public curious about mathematical tidbits, and for mathematics students and faculty especially from India and the US. They discuss curricular issues, histories, cross-cultural perspectives, and typical faculty and student scenarios extracted from classrooms, professional meetings, conventions, and the department hallways.

How did the *Reflections* start?
'Serious' writing has been my nature for many years. In nutshell, *Reflections* are re-incarnations of old-fashioned letters that I loved to write, receive, read, and collect since my 20s. Seven years ago, after having written several *Reflections*, I looked back at them for a possible categorization. It then became essential to separate out the *Mathematical Reflections*, as there is something technical and alarming about mathematics in the minds of laymen.

Is there any order?
No. *Reflections*, by nature, bounce from one theme to another. Most of them do start off with 'local' incidents, but expand to 'global' levels. Two or three sub-themes or sub-topics are packed in each one. My writing is, indeed, compressed - lean and fat free! Any *Reflection* can be expanded into several pages, though requiring perseverance to pull out a yarn. I do not have the temperament to stick with one theme, no matter how intensely my mind is locked in at the outset. Passions, by their very nature, rise and fall quickly.

Is there any chronology of *Reflections*?
I wrestled with chronology for a while. Eventually, I decided in favor of the dates when the *Reflections* were written. It gives a sense of history to the entire collection. At this stage of life, I like history in every shape and form. Besides, there is always an incident, big or small, that triggers a particular story. The date, in the parentheses, at end of each *Reflection*, is the starting date. Usually a *Reflection* is finished and clicked away within 48 hours - of course, after several readings and hard copies.

Over the years, I have developed a mailing list of nearly 200 readers including faculty, graduate students, friends, and former students in the US and India. In one respect, these **Reflections** have been tested over a good statistical sample. I do not have a website yet! A student did create a blog for me, namely: www.reflectiongalore.blogspot.com. But it is sparingly used. I am so conditioned to copy-paste, and bcc e- mailing. Keeping abreast with technology is ultimate distraction to my writings.

Was there any selection?

Any classification, based on topics, is amorphous and not worth it. Moreover, it upsets the time line. I have not taken any **Reflection** out! It is like a mother who loves all her kids. However, while editing, I questioned the educational and informational merits of each **Reflection**. Each and every **Reflection** has passed this test! This may vary with **Reflections**. Some mathematics articles, molded into **Reflections**, are 20+ years old. It is like the silverware - lying in a cabinet for months, then taken out and polished to bring luster before setting them up on a table. In some cases, the revisions were extensive, but the core has remained undisturbed in each case. However, this Volume I contains the first 98 **Mathematical Reflections** written through the end of 2006.

Two words on my English

This **reflective** preface won't be complete without it. I went to high school and college in Bathinda, the hinterland of Punjab. I was a rebel when it came to learning grammar and spellings. For instance, I spelled 'military' as 'miltary' for years, and got penalized. In fairness, neither the punctuation was taught well, nor was I receptive to it. For years, I have believed that the contents should prevail over the dressing of the grammar. Punctuation, like black pepper, should be sprinkled over a dish for its flavor, not for the taste.

Well, once I started having ambitions of becoming a non-fiction writer of stature, I eventually realized the importance of grammar and punctuation. At least, clear sense and meanings in writings must be conveyed. In addition to these pronouncements, I enjoy coining new expressions and words, like '**matherticles**' in the title of this volume, derived from 'mathematics' and 'particles'. Minor grammatical rules are flouted here and there, particularly in the usages of the article, 'the', commas (,), semi colons (;) and dashes (-). Also, there are nearly 20 common words like – maybe, sometime, anyone, anything, forever, likewise - spelled one word in the US, but in Bathinda, I learnt them as two. However, my usage is mixed, and thus fair. Like if - then structure of a sentence, there is though-but structure, which is un-American English.

However, there is a real story about my English accent and comprehension. Forty-two years ago, when I first faced the accent issue in the US, I fought it long to retain my North Indian enunciation. Coming to the US at 29, my vocal cords and guttural muscles were set. Mathematics being a 'language', I focused on it for my doctorate too. Now I teasingly tell my students, "You are learning from me an English accent free of cost in addition to mathematics that you pay for. Moreover, nearly 300 million people in North Indian subcontinent speak with my accent." As far as comprehension is concerned, Indiana University Mathematics Department had laid out a policy that for Indian students, English was not their foreign language, as they were better than Americans when it came to reading and writing. In order to satisfy the two-foreign language requirement for my math PhD, I had to choose the two out of French, German and Russian.

Indians have been producing great literary works in English for the last 100 years, but they are not recognized in academic circles, or for international awards, simply because their literary style is different from the British and American norms. That has bothered me for years. Now that I am 70, I can confidently say that my writing style neither falls under standard English of the US, nor of contemporary India. It is uniquely Indo-USA in usage, imagery, structure and phrases. It is integral to my being, and hence carries a distinct stamp on my writings.

Dedication
When it comes to motivation and inspiration for writing these *Reflections*, it is totally internal. The urge to do it had been bubbling and building up for years. The more I wrote, the bolder, clearer,

and in some sense, more courageous I became. Writing for public consumption is actually uncovering one's soul. It is very different from a personal journal and diary. After all, who wants to read a tired and tried yarn? Several readers have pressed me to put these *Reflections* together in books.

Besides *Mathematical Reflections*, I have written nearly 700 general *Reflections* on various subjects of human drama. In mathematical jargon, the range of topics goes from minus infinity to plus infinity. My granddaughter Anjali, whom I call *Sherni* (Lioness), has been associated with most of them as a reader, critic, editor and PC file manager. In appreciation of her longest association, this volume is dedicated to her. Incidentally, mathematics comes easy to her, but she majored in Cognitive Science, a highly mathematical interdisciplinary field. Last April, she finished her BS from the University of California at San Diego. She alone has 'appeared' six times in my general *Reflections*.

Thank-Yous

The two persons who have initially helped me to develop in this direction are Renato Estacio and Scott MacDonald. Both were my students in math courses taken 7-10 years ago. They are now good friends. Renato, once Director of UNLV's Writing Center, gave many tips on 1-1 basis, and through e-mails. He is starting on his PhD in Communication. Scott is already a published author. He once owned a computer store, but sold it to become a full-time student. Subsequently, he earned BA in math, and now pursuing PhD in Mathematics Education from UNLV. Both of them red inked several *Reflections*, and I studied my errors for days. Scott used his computer expertise in putting together the first draft for the pre-publishable form of the book. It was a new learning experience.

There is a grain of truth in the saying, "the old dogs do not learn new tricks". A plateau hit me in my writing progression, and I rebelled again, under the excuse of establishing my unique style. In August 2007, when I decided to sign the contract with Trafford Publishing, my son called, "Dad, this being your first book, are you completely

satisfied with the final draft before sending it out for publication?"
Well, I had assumed that editing, as done by my granddaughter and
her friend, was enough. At times, enough is not enough.

During my sabbatical leave from UNLV in fall 2007, I spent some
time with Ishwar Chandra, a friend of 20 years, settled in Chandigarh,
India. Having been a publisher, editor, and printer, he understands
all the nuances of industry from inception of writing to publication.
In fact, it was the literary beauty of his small pamphlet that got us
acquainted a year before we actually met. He is incredibly demanding
in perfection. He convinced me to go back to my *Reflections*.

Conclusion

Re-reading and editing of the *Reflections*, indeed, was rewarding,
as it turned out to be a new journey into my inner recesses. Since
I literally erase a *Reflection* as soon as a new one takes possession
of mind, I found new delights in old *Reflections*. It certainly slows
down the writing of new *Reflections*. But the rewards in realizing
depth and clarity of thoughts are immense and lasting.

During this period, I also realized that perfection is only a state of
mind. There is no finality about writing a piece. Any time I read a
Reflection, say for the 10th time, after a few weeks, months, or years, I
always find something to 'improve' upon it. Mathematically speaking,
it means that one can spend the entire life in improving upon one
Reflection alone. It is another version of ancient **Zeno's Paradox**, just
shifted from no physical motion to no new writing!

Writing has offered me a glimpse into *Self-Realization*. After a
while, you have to let the contents prevail. In life, you can't to be too
protective of your nascent ideas, as we are towards our children. Both
have to face the scrutiny of the world outside. Hence, it is time to
look forward to the next phase of how these *Reflections* fare, or how
well prepared they went out.

In conclusion, any 'errors' are wholly mine. I would appreciate receiving any feedback via e-mails at: bhatnaga@unlv.nevada.edu. They will be gratefully acknowledged.

Satish C. Bhatnagar
Sept 12, 2010

TABLE OF REFLECTIONS

ON COLLEGIALITY IN MATHEMATICS
(A Reflective Note to the Ad Hoc College Committee on Math Department)

While pondering over collegiality amongst math faculty, it occurred to me that **by and large the discipline of mathematics does not breed collegiality**! I remember my IU days where the world famous mathematicians, P. R. Halmos and R. P. Gilbert, having verbal brawls, and never greeting each other in the hallways. Mathematics is a one man show; apart from free loaders and riders on research papers, particularly in small institutions!

On the contrary, in experimental sciences, a person, constantly working with at least a couple of people in a lab, develops some interpersonal skills. Let this be a boundary condition as you deliberate on your recommendations. First, I appreciate you all on this Committee for helping the Dean and the College so that the Math faculty and Dept Head move in a right direction (The path of steepest ascent is always short, but telling).

Yes, there is a growing lack of collegiality and I agree with Dean's decision of holding off the implementation of much awaited Math PhD program. However, this is precipitated by only a couple of individuals to oust the Head. They never accepted him. Some initial support was given on a belief that he would quit. In Oct 2002, he came close to it, but he has now regrouped. Last year's storm has indeed toughened him up.

It is not out of place to add that continuous challenge to the authority of the Head, and attitude of non-cooperation is partly due to the higher administration. I am told, that after the loss of three law suits costing the University a million dollar, the Chair would have been fired in any another institution. UNLV did not even reprimand him! A perception exists that the Administration is afraid of law suits! Thus it has emboldened a few faculty members to challenge the authority. Any lack of collegiality is not as wide as it looks.

Recent promotions to professorship and administrative appointments suggest that it pays to serve on the Senate and NFA. No wonder recent problems are being caused by faculty on Senate and NFA. Also, the dean's office has not been stable. Since 1995, no dean has served the College for more than two years. It has hurt the College and Math Dept in particular. The Y-2000 decision to split the Dept in two divisions without meeting the faculty was not well thought out. In 2002, the new Head deftly put the Dept together, and the Administration rightly supported him.

Lack of collegiality does raise its ugly head at the time of hiring, and lately in selecting graduate assistants. There were two failed searches this year. To eliminate it, the next 4-5 hires may be done like it was done to hire Phanord. Nearly 50 % of the faculty, being Asian, comes from non-democratic societies and tribal cultures. 75 % is over 50 years and over 40% are full professors. Most hires are in the areas of pure mathematics like number theory, logic and topology that are not in national trends, and much less at UNLV.

Hiring of 2-3 non-Asian associate/full professors with a successful record of guiding PhD dissertations and acquiring grants will change the atmosphere. For enhancing research and programmatic collaboration with other departments and colleges, I venture to suggest single/joint hires in the areas of Bio-Mathematics, Mathematical Physics/Chemistry, Bio-Statistics, Mathematical Seismology, Mathematics Education, and high speed Computational Mathematics.

The Head is now at a cusp of understanding and leading the Dept. But he needs the Administration's support both in his decisions and resources at least for two more years. Early on, he drew from his experience of a small (Carnegie Master I) Whitewater campus of University of Wisconsin System. To foster faculty governance, he held too many and too long Dept meetings. They proved counter productive. Often the discussions turned hot and went out of control.

Being at UNLV for 30 years now, I have not known a department split in two for lack of collegiality. There are horror stories in some depts., like Economics for one, but they are still intact. I have always believed in one Mathematical Science Department. If I had my way 20 years ago, I would not have let four computer science faculty members leave Math Dept and join Electrical Engineering!

Any move in Math Dept of taking out 4-5 faculty members to the proposed School of Science and Mathematics Education has the potential of a Class Action law suit that the University would not afford to lose. However, placing 1-2 in the School, or assigning them to other colleges may be effective.

(Aug 01, 2004)

My Tale Of Two Professors

This year, I served on the ***Distinguished Teaching Award Committee*** of the College of Science and Mathematics. The only available data were the nominations by the students. There was no feedback from the department chairs or faculty, though solicited. Students, mostly undergraduates, nominated professors currently teaching them, or had them in a preceding term. The Committee also took into consideration any cumulative record of past nominations.

My thoughts went back to my four years of bachelor's in Bathinda (India) when I took all math courses only from **two** professors; Deva and Bansal. Bansal had no problem in teaching for the first two years. For getting Honors in Mathematics, one had to pass seven papers (US equivalent to 3 one-year sequences) in one shot at the end of the fourth year. Deva taught five of them! Bansal could not handle his share due to personal problems and difficulty of the material.

Invariably, Bansal would run in late by ten minutes and put out his cigarette at the entrance of the classroom! After writing out one or two examples already worked out in the text, he would walk out five minutes before the 40-minute period was over. Deva was just the opposite; extremely conscientious, thoroughly prepared, organized, and always challenged the students.

Under the then prevalent system, the final exams (each 3-hour duration) were not conducted and written by the instructors. They were set by examiners from other states and proctored by yet another group. As the finals drew closer, the panic increased in Bansal's papers. Complaining to the principal made it worse. Dropping a paper meant delaying graduation by one year. The alternatives were to pay for tutors, or study with self-help books. Often, Deva helped out individual students in Bansal's papers too. With perseverance and hard work, my state of frustration slowly changed into self-reliance and confidence. It continues to pay me off.

I became aware of this inner transformation 12 years later while working on my PhD dissertation at Indiana University (USA). Within a few weeks of choosing the area of dissertation, my supervisor suddenly decided to leave for Germany for one year. He gave me the options to come along, change supervisors, or do the work independently. I stayed back for my family. Under this pressure, not only I was able to find some research problems, but also solve them.

During one of those moments I felt my gratitude welling up for Bansal. It took fifteen years! I don't mean to imply that the lousiness and incompetence of an instructor motivate the students into studies. Never! Rather, I discern the impact of Deva and Bansal on me lasting and valuable. Recognizing excellence in teaching at the college level is a healthy sign in the US education system. A caution has to be exercised, that the selection process does not turn into a popularity contest amongst the students. **The impact of a good teacher is indeed life long.**

(April 05, 1988/Revised: Nov 16, 2006)

Epilogue

Bansal died 10 years ago. Deva at 75 enjoys commercial tutoring on a part time basis after retirement. His marrying my classmate and his student, in 1959, was a story of its time!

Prefatorial Remarks

The following article was written twenty years ago. The entire culture of classroom teaching has changed since then. I continued holding review sessions for a couple of years, and then stopped it. Reason, hardly any student came for extra help, though free. Also, UNLV students are too busy at their jobs, carrying overload of courses, and some raising families. Certainly, Las Vegas factors into every aspect of life.

Was I ever given any consideration at a time of merit and promotion? No! In a university setting, any time away from research is 'damaging' for professional rewards. However, one must live in idealism for a while. Currently, the Department offers free tutorial services, but not more than 2-3 students, in a class of 40 College Algebra students, take advantage of it during a semester!

A Perspective On Review Session

In the lower division mathematics courses (100-200 levels), review session is a must. The number one question is how to schedule such sessions, when the syllabus is so tight that one hardly has the time to finish it. In a 3-credit course, three hourly tests and a comprehensive final are given. There is no way that at least four days can be carved out of instruction for reviewing.

A solution to this problem is to schedule the problem sessions at non-class times, and run them of two hours. This forces a discipline on the students. If you are motivated to learn mathematics then you better take time off, if necessary. If the instructor is willing to come for additional hours, then it behooves that students take advantage of this opportunity. Nearly 70-80% students attend it. Those who don't come usually don't need extra help at that time.

The next question is how the reviewing is done. It is not repeating the lectures. During the review sessions, students take the initiative in the sense that a particular problem is taken up, provided he/she has actually attempted it, but failed to get it. The students must bring their work sheets of the problems before seeking help. You don't learn mathematics by watching some one do it! I don't hesitate to discuss a problem even if it has been already included in the test. If it happens, then unknowingly, they have hit a jackpot! Problems are invariably solved by participation, thus benefiting everyone present. Such review sessions convey integrative view of the topic.

Letting a student ask any question from anywhere builds students' confidence in the instructor. It is essential in their long term learning. Since problem sessions last up to two hours, there is enough time to reach out each student's problems at least twice. This individualization helps in removing the inherent anxiety, which eventually translates into better feeling for the subject. There is a significant correlation between the scores in the exams and participation in review sessions.

There is a corollary to this kind of reviewing. Often the general performance over a certain material is not satisfactory. Since grading is not on curve, students are given supplementary tests, but not during the regular class time. They are scheduled on a Sunday morning. The idea is to force them to study over the weekends and prioritize between their studies and jobs, if any.

An objection to this approach is that why an instructor should go to this length and put in extra time and effort. I believe that teaching does not stop after the regular class hour is over. It continues outside the class as well. Helping students learn and appreciate mathematics is as challenging as solving some research problems. While working on a research problem, one seldom shuts out one's mind after working for an hour or two on it. Instead, it goes on. Recent reports on the US students, not doing well in mathematics as compared with other countries, totally fail to point out the difference in the quality of teaching between the US and other countries. Finding extra time for students is commitment to teaching. That alone secures the future of the discipline in the hands of the next generation.

(Jan 1988/Revised: Aug 2008)

My Teaching Philosophy And Methods

Since 1961, college teaching has been my sole profession. I try to blend the strength of my best teachers, and avoid the weakness of others. The teaching profession from the stand point of the instructor rests upon three main pillars, 3C's: Competency, Communication, and Compassion. Graduate work in mathematical physics and mathematical seismology done in India, and in computer science and mathematics at Indiana University, has given me a tremendous edge in relating mathematical ideas. Combining it with vitality means everyday vigor and freshness. It comes from rotating a course after 2-3 years, choosing different textbooks, and designing new courses etc. As a consequence, I have taught 38 regular courses and 12 experimental courses.

Mathematical is unique in one sense that the traditional research done at the doctorate level, cannot be easily brought in at the undergraduate level. However, this dichotomy between teaching and research in mathematics gets thinner as the level increases. Teaching, not periodically watered and fertilized by new ideas, becomes dry and stale. Research activity which in mathematics easily gets distanced from students only becomes egocentric.

The three C's set a conscientious teacher from the rest. While presenting the material, I can't stand to see anyone lost during lectures, which are highly participatory. The smart ones are encouraged to do challenging problems and earn extra credits. It takes me 2-3 weeks to memorize students' names and their academic interests. Students at UNLV are very non-stereotyped! My best class hour is when I feel like a musical conductor, having an eye to eye contact with every student. It is essential to build students' confidence in my ability to teach the course and theirs in finishing it. It results in minimum dropout in the courses; in fact, very little after the third week.

I practice a common principle of psychology that students learn at different times and rates, by giving them additional opportunities to demonstrate their mastery over a particular material. Review sessions are held at times different from the regular class periods and supplementary tests are scheduled on Sunday mornings. At any given time, during the semester, a student is fully aware of his/her grade. Yet, he/she has a shot for the highest possible grade in the course on the basis of the final comprehensive exam. Until the final exam is over, students are mastering the material. I never curve my grade distribution.

A corollary to this individualized approach to teaching is that the human factor begins to grow as the semester rolls on. If a student misses a class, then I don't hesitate to pick up phone and check it out. The net effect is a quantum jump in the performance. For me this concern for the students is natural, and partly it comes from the culture in which I spent my college days.

(Feb 07, 1989)

Note: I have changed little in 20 years, except holding review sessions on weekends.
(Oct, 2008)

Prefatorial Remarks
The following piece is not one of my typical *Mathematical Reflections*. It is not an article, or paper written for any mathematics journal. It had a limited circulation. However, the thoughts put together twenty years ago, have stood the test of time. They are truer today, and shall be at any place and time, where research carries premium over teaching. Such a work gives a great inner satisfaction in life.

THE RISE OF BERNOULIS AND HOPITALS

This semester while teaching introductory calculus, the following footnote on L'Hopital as given on page 359 of the text [1] caught my attention!

"L'Hopital's Rule should actually be called "Bernoulli's Rule," because it appears in correspondence from Johann Bernoulli to L'Hopital. But L'Hopital and Bernoulli had made an agreement under which L'Hopital paid Bernoulli a monthly fee for solutions to certain problems, with the understanding that Bernoulli would tell no one of the arrangement. As a result the rule described in Theorem 6.14 first appeared publicly in L'Hopital's 1696 treatise, *Analyze des infiniment petits pour l'intelligence des lignes courbes*. It was only recently discovered that the rule, its proof, and relevant examples all appeared in a 1694 letter from Bernoulli to L'Hopital."

It struck me that this shady aspect of research in mathematics has gotten worse ever since! What perhaps existed in isolated individual cases has now institutional and administrative support at every level. Reason, the colleges and universities have put their entire premium on research, which is only recognized by traditional published work. No other form of creative and scholarly activity earns any personnel consideration. Two of the commonest cases of research blackmail and pollution (What else to call it, Research AIDS?!) are the following:

As soon as a new member joins the faculty, the specter of tenure starts hovering over. The system mainly works on voting in various committees. Therefore, 'research collaboration' is encouraged which means putting the names of some senior faculty as co-authors on papers to be published, or presented in professional meetings.

The worst known cases are the so called 'foreigners' with U.S. Ph.D. degrees. Many of them don't get tenured track positions easily. On visiting positions, they go from one institution to the other every year or two. As new faculty members, some would have no qualms in putting the names of persons who have little interest and background in their research, if that would assure them tenure, or a tenured track position.

It is obvious why such men of intellect resort to it. The tangible rewards are in tenure, promotion, sabbatical leave, merit raises, and no less, the opportunities to travel to places all over the US and abroad. They are all tied with research, though it is one of the three components of the academic profession; the other two being teaching and service. The impact of such a research environment is very negative on quality teaching and students' attitudes. The grades become inflated and syllabi are not properly covered. As an offshoot, the students start expecting soft teaching from other instructors.

At this point of my analysis, let me note that all that has been observed is generally prevalent in mathematics departments of non-PhD. Institutions like University of Nevada Las Vegas (UNLV). That is the focus of this article. These institutions are not sure of their identities. They always look up at established 'big' universities, and often make wasteful efforts to clone themselves. One of the reasons is that faculty members are seldom able to snap their umbilical chords from research conditioning of their graduate days. Realities of the employer institution with respect to the local population and geographical settings are completely sidelined.

Lynn Arthur Steen, former president of MAA (1985-87) in a series of articles [2, 3, 4] effectively brought out, that scholarly expectations in undergraduate mathematics are very different. However, without any organizational and follow-up actions, the status quo is not going to change.

The *Notices* of December, 87 and February, 88 contain the data on the number of math faculty and the number of those faculties who have published a technical paper or a book during the last three years (1984-87). The percentages of 'publishing' faculty, in mathematics from ten leading universities reporting the data, are noted in the parentheses:

1. Berkeley (80)   2. Harvard (90)   3. Minnesota (74)
4. Courant Institute (80)   5. Ohio State (66)   6. Michigan St. (72)   7. Pennsylvania (88)   8. Texas at Austin (56)   9 . Wisconsin (86)   10. Arizona (78).

At the other end are ten small non-PhD. institutions. Their size was determined by two parameters, that during 1986-87, the number of granted bachelors degrees is less than ten, and that of masters degrees less than or equal to three. The percentages of their 'publishing' faculty are similarly noted:

1. UNLV (53) 2. Northern Arizona (41) 3. U Colo. Colorado Springs (100) 4. U South. Miss. (56) 5. U Nevada Reno (39) 6. Seton Hall (50) 7. Baylor (56) 8. James Madison (48) 9. Hofstra (70) 10. George Mason (72).

Another related set of data is from the number of papers presented at the Atlanta meeting as reported in the January, 88 issue of the *Notices*. It has two components. One is the noticeable increase in the multiplicity of the co-authors of a paper, going up to four in some cases. Second is, that the number of papers presented by the faculty in the first group is 36, against 14 in the second group.

Before jumping to any conclusions coming out of this data, we have to be cognizant of the fact that whereas the percentage of non-PhD. faculty holding even a half time position is nil in the first group, it can go up to 30% in the second group. Also, the average teaching load/faculty/term in the second group is at least twice that of in the first group. Now, it is for the readers to make their own interpretations. Personally, I tend to agree with the observations made by Steen [1, 2]:

"For too long, mathematics and mathematics teachers have suffered from a rigid, narrow definition of professional activity. To save face with our peers in other sciences and the humanities, we demand of ourselves a productive research program; to save face with our peers in mathematics, we adopt the mathematician's elite definition of research. The result is often confusion, frustration, and well-intentioned hypocrisy in tenure and promotion reviews".

Acknowledgement: Thanks to Tom Schaffter for his suggestions and support.

References
1. Ellis and Gulick, *Calculus* 3rd ed. Harcourt Brace Jovanovich, 86.
2. Steen, Lynn Arthur, *"Crisis and Renewal in College Mathematics"*, *Point of View*,
    The Chronicle of Higher Education, July 24, 85.
3. ----------, *"Renewing Undergraduate Mathematics"*, Notices, Aug, 85.
4. ----------, *"Restoring Scholarship to collegiate Mathematics"*, Focus, Vol 6(1), 86.

(May, 1988/Revised, Aug 2008)

IMPORTANCE OF
MATHEMATICAL EXPRESSION

In lower division courses (100-200 levels), there are two kinds of obstacles that the students encounter in learning mathematics. One is due to the nature of the subject itself; its rather abstract concepts, definitions, notations, emphasis on drill problems, and lack of relatable applications. The other is simply due to actual doing of the problems. In no other discipline, the writing has such a great correlation with comprehension, as in mathematics.

The students and general public alike are heard making remarks on how mathematics is difficult, how it freezes them, and how they hate it. By watching individual students doing problems right in my office and classrooms, I have observed that often mathematics has very little to do with their difficulties. If it is a word problem, then in eight out of ten cases, students have not fully comprehended the text of the problem. Thus, the difficulty belongs to the domain of English 101 or 102 courses! Invariably, the students are making avoidable and unforced errors due to their weird writing styles. When pointed out, they are amazed at the source of their difficulties.

What is, after all, **Mathematical Expression** (ME)? It is a combination of crisp literary phrases, mathematical symbols, abbreviations, and their systematic manipulation. An analogy with computer programming is apt. **What documentation is to programming, mathematical expression is to the doing of mathematics**. Without documentation, a programming code becomes alien to its author after a little while. Of course, it always frustrates a third person.

When to emphasize upon ME? A touch of curriculum and cultural history fits in here. By and large, India has been influenced by the British system of education. During my era of 1950's, as soon as algebra was introduced in high schools, students were constantly reminded that without ME, full credits won't be earned for the solutions of problems. It used to irritate the students, as what more is there besides arriving at the correct answer?

Learning is slow and often students learn by watching. If one looks at the mathematics textbooks written before the fifties, then one would find a marked emphasis on the art of problem solving. Initially, ME may appear redundant, but it eventually makes mathematics easier. The study of Euclidean Geometry, introduced in the ninth grade, made the acceptance of ME natural. I believe that nothing surpasses classical Euclidean Geometry when it comes to exposing the students to deductive reasoning, structure, and beauty of mathematics, at elementary level.

Today, situation of doing mathematics in schools is dominated by scantrons and right/wrong answer sheets. What needed are descriptive solutions to balance the current practices. There is no reason for the colleges to wait and see changes take place in schools first. The instructors in colleges can initiate and take the charge of stressing ME in all lower division courses.

My approach has been to encourage the students by giving extras credits for ME. It is integrated into every course. Once the students have realized the advantage of clarity and good performance, they never go back to their sloppy styles. Here are the seven rules of ME:

Seven Rules of Mathematical Expression

1. Starting from the end of the left margin of a worksheet, write horizontally to the right, as far as necessary (not, as far as you can go!). Avoid breaking a 'mathematical sentence' in two lines. Under no circumstances, divide a standard work sheet in different vertical columns. Limiting space for the solution of a problem, limits the thinking process. That is inimical to the intellectual development. Those who save money on paper, while doing mathematics, pay in lost credits!

2. All major mathematical steps should follow a vertical down-flow. Mathematical solution is a cascade of logical steps. A short 'implication' may be continued horizontally.

3. A corollary to 1 and 2 above is that a solution of a problem should never meander into spaces inside the left and top margins. Currently, there is disregard to these two points mentioned above. Often a solution appears to be absolutely zigzag, circuitry, or looking like a doodling noodle!

Necessary mathematical symbols may be used at the beginning of each line, if necessary. For example, the symbols, $\therefore$ for 'therefore', $\rightarrow$ for 'it follows' / 'it implies', $\because$ for 'because' etc. make writing look compact. This also brings sharpness in thoughts.

4. Neatness in work always breeds clear thinking, is a popular phrase. It is all the more true with students who are taking their first math courses. The excessive use of the erasers should be discouraged, as it messes with the work space, and waste precious time, particularly during tests. Working under stress, for any reasons, coupled with erasing and over-writing, I have known students making weird errors. There is something to be learnt from fresh mistakes! If they are quickly erased, the errors are likely to be repeated. Instead of erasing, put a gentle cross over a wrong solution, and continue to work in a clean space, or use the back side of the worksheet.

5. State in words, abbreviated, if necessary, what is to be executed in the next step? For example, 'Multiplying both sides by (-2)', 'Subtracting Equation #2 from Equation #1', and clearly identifying the variables for what they stand for etc. This creates an awakening effect and general alertness for the work to follow. Otherwise, for instance, students generally forget to reverse the direction of the inequality in case division by a negative number, or they may add a few terms, or partly subtract Equation #1 from Equation #2. With ME, at the end, it makes the reviewing easier. Errors are more likely to be detected.

6. If I have to put my finger at one single source of big mathematical errors in all lower division courses, then it is the use of simple parentheses, braces and brackets in mathematical work. To emphasize their importance, I frequently remind the students that parentheses do not cost anything, but save lot of points.

7. The scratch work should not be mixed with the solution to be graded. At the same time, scratch space or sheet should not be inconveniently located, that going back and forth may cause errors of coordination. It should be clearly identified with respect to each problem.

(June 1988/Revised Oct 2008)

BOXING AND MATHEMATICS

Last Saturday, my brother-in-law invited me to watch the super welter weight (154 Lbs. division) tile fight between the champions; Fernando Varga (WBA) and Oscar De la Hoya (WBC). By the time I arrived, the last undercard fight was in the middle. I fixed a drink and a plate of BBQ chicken and burger, and I sank into a sofa chair before a huge TV. There were a few friends of my nephew too, who were excited about the fight and were yelling for one champion or the other. When they asked me about my favorite, I said, "none; I just love the dynamics of fight."

You have to grow with boxers, singers, and music bands to develop a like or dislike for them. It is the same in mathematics. If for example, you have not grown with powerful graphic calculators, PC's, taken courses, or stayed updated with new areas of mathematics, most often one is 'left behind' in the sense as being out of fashion. However, I know a great mathematician, Hans Raj Gupta who knew little about modern areas of mathematics viz., algebra, topology, logic, statistics, and even analysis. Besides classical geometries and mechanics, all he knew was pure Number Theory. But, he knew it better than any one in his era.

All through the fight, my thoughts were going back and forth between the fierce actions of the fight and my efforts to see comparisons and contrasts between mathematics and boxing. Here are my raw thoughts put in black and white. There is no special ordering, or in the number of points.

1. The most striking common feature between math and boxing is that both are for the **young at heart.** It is literal in the sense that just as no great theorem is proved after one turns 35, so is the case with boxing. Boxers go into retirement after 35.

2. Just as **men dominate** boxing, so they do in mathematics. Of course, there are efforts to promote women's boxing, but it is like a few women in mathematics who are there due to some family

support or strong individuality. It is due to women's different priorities in life.

3. Boxing is the ultimate **one-man sport**, so is mathematics at its stage. Since mathematics is getting pervasive in applications in various fields of engineering and social studies, and natural sciences, the applied aspect of mathematics is heavily dependent on collaborative efforts. By and large, all great theorems in the history of mathematics are the results of individuals alone.

4. Before a fight, the fighters rigorously **train in isolation** away from their wives and girl friends for months. They also train to focus the mind. In my knowledge, wives of mathematicians are not very much enamored of the routines of their spouses.

5. During the fight the boxer has to have **very sharp eyes and reflexes,** otherwise in a lapse of a split second, the head maybe knocked off. While solving a math problem the intensity of concentration is highest. Also, a student of math must be fully aware of what is given as a part of a problem, and should be able to apply every thing in the pursuit of its solution.

6. **Boxing draws blood**. The bloodier is the fight, the more fans enjoy it and get their money' worth. After the fight, De la Hoya put it in perspective when he said," When I draw blood, then I like to draw more."

7. **Money flows** around a boxing flight. After the flight, the high rollers and common visitors in Las Vegas leave more money at the tables of casino games. There is a greater partying after the fight. The limousines have a great business. The brothels in Pahrump, 50 miles away from Las Vegas, have significantly higher business after a fight.

8. **Boxers come from tough inner neighborhoods**. It is therefore dominated by Afro-Americans in USA. or Latin Americans. It is the gateway out of poverty for them. If I may use the popular

term nerds for kids obsessed with math and computers, they rarely come from poor sections of the society. Most of them come from families of above average means. Those who go for the study of mathematics make a conscious decision of living in poverty by choice. What a contrast!

9. Great mathematicians, who try to prove unsolved problems or pursue problems at the cutting edge, are eventually **burnt out**. It may be due to short circuiting of the neural network fused by hours of concentration. Most boxers pay a heavy toll in later years. The impact of blows to the heads overheats the cerebellum. It shows up in later years in various health problems.

10. **Both boxing and mathematics are measures of a nation.** Today, the US is Number One in mathematical productivity as well as in boxing achievements. The best boxers of foreign countries come to USA for training and become professionals.

11. Watch little kids playing in fields and working in classrooms. They display their innate love for early boxing the way they attack their playmate in a sudden rage! It is boxing in the most uncultivated form. In the classrooms their love and grasp of number concepts is evident. The kids learn to count numbers far earlier than they recognize alphabets.

12. The urbane parents discourage kids from becoming professional fighters, as they do to kids becoming mathematicians. **Particularly girls in US are never encouraged to pursue math**.

13. Boxing is fought in 17 divisions from the lowest division of Straw Weight at 105 Lbs. to the Heavyweight divisions of over 190 Lbs. The two consecutive divisions are apart from each other by 3-15 Lbs. I think it is more appropriate to compare divisions with some 20 years of school and college, where the kids learn and play with math problems depending on their needs and potential. Roughly twenty years include 12 years of high school, 4 years at bachelors, and at least four years of formal graduate

course work. Boxing skill and stake increase, as one moves from a division to higher divisions. Just as not every one makes to the top division, so does it happen in mathematical achievements. Mathematics also changes from complete utilitarian and applied to abstract and conceptual.

14. Let us look at the **making of professional boxers and professional mathematicians**. Most professional boxers learn their art on streets, and often from the penitentiaries and prisons. Rarely one picks up math from the street corners. It is cultivated, though it may have years of bursts in early graduation in high schools and colleges. Mathematicians charge their batteries on sabbatical leave. Do the boxers hone their skills in prisons? Now that the survival schools in the US are gaining acceptance, then the odds of growing to be 18 in some inner city neighborhood is no more than graduating from a survival school.

15. **The aura of boxe**r in a school and college day is probably the envy of a majority of the 'lesser' ones. However, no one is envious of a math whiz. On the contrary, he may be an object of some ridicule.

16. The **financial awards** in boxing are fabulous even when compared with superstar mathematicians of the world. In this fight, Hoya is expected to earn 20 millions; his assets are worth 100 millions. And he is not even 30 yet! In mathematics, I advise every one that no matter how much you love math, you must have a minor or additional major in some other area that may guarantee a moderate life style. A pure math PhD does not garner $40,000.

17. **Both boxer and mathematicians are there for the love of what they do.** When asked what motivates him to fight with 100 millions in banks and beautiful family, Hoya aptly put it --the passion for boxing. It is for the love of doing and teaching mathematics that most math professionals choose to work despite living in a relative poverty for the rest of their lives.

18. **The legends in boxing and mathematics**. It is not just in the names of persons who touched greatness in their domains. This fight was billed as Bad Blood. I remember fights labeled **Rumble in the Jungle** and **Thrilla in Manila**, and so on. Each major fight is appropriately named to capture its aura both for marketing as well as for boxing business. Mathematics has famous theorems like the **Fundamental Theorems of Arithmetic, Algebra and Calculus**, or after the names of persons like **Pythagorean Theorem**. Names after theorems have to stand the test of time. Otherwise, the names just fall off into oblivion after a few years.

19. A striking similarity between math and boxing is the intense desire for a single knock punch. In high schools, the prevalent impression is shorter the solution better it is. A boxer is feared when the opponent is knocked out in the first or second round. Just as various boxing skills, stamina and ability to overcome a bad round are all tested, if the fight goes to the last round, so is it in math too where a long solution brings various techniques and ingenuity.

20. Another similarity between math and boxing is the practice and training. One has to do scores of problems to gain insights into a topic. Boxers, in ring with all protective gears, spar with partners for several rounds each day. It is in boxing that a man comes close to being killed with blows. Therefore, the boxers have to be very sharp. On the other hand, mathematics provides the ultimate calisthenics to the human brain. That is why though mathematicians are sitting most of the time they are able to burn a lot of calories because of overactive of brain cells while working on tough problems. One day, I asked my son majoring in math, if he ever spent a couple of days working on a single problem?

21. Techniques of solving a math problem are similar to the approach that a boxer takes. A boxer uses psychology to measure up the opponent, say by staring into the eyes. A problem has to be fully comprehended before assaulted. Initially, a solver dances around the problem by plugging numbers, checking a few special cases,

and trying a special case. In boxing, that is like jabbing, attacking and retracting. Often a point comes when there is no headway and one wants to pull one's hair. That is like taking blows when pushed on the rope. It wears out the attacker, if one remains under control. Upper cut and power punches to the body wobble and weaken the opponent as a problem is cuts and handled in pieces before a full attack is launched.

22. There is a strong similarity between mathematics and boxing when it comes to competitions. Mathematics is the only discipline where problem-solving competitions are organized in every form. They are refined at junior/HS and College levels, and go from local to all the way to nationals. It further goes to international called Mathematics Olympiad. 88 countries participated in the last Olympiad where individual and team problem skills are tested. Of course, in boxing, the competitions have more structures outside the confines of schools and colleges.

23. More to come!

Conclusion

The purpose of the whole exercise is that no human activity exists in isolation because, they all stem from one human being. And hence bridges can be established between any two areas, particularly between math, my lifetime pursuit and any other discipline, or endeavor, be that intellectual, physical, or spiritual. An integral and holistic approach taken by an instructor and student during undergraduate years is likely to go a long way in finding meaningful relationships of formal education with life at large, later on.

(Sep 18, 2002)

Mathematics And Theatrical Arts

The following outlines on mathematics and acting provided a backdrop for the freshmen taking the first part, **Honors Mathematics I** (HON 140H) of the required two-semester sequence. They earned extra credits by making mathematical connections with play acting after watching a bunch of them on campus; free for the students. It was not required assignment.

1. **Memory**: It is perhaps much more vital in plays than in math courses. Just ask any theater major, or watch performers in their roles.
2. **Practice**: For hours and hours, actors practice alone or in groups!
3. **Integration**: The oneness with a math problem matters in solving it, just like, for a while, an actor fully identifies with the character he/she is portraying. Actors are like math variables x, y, and z that take different values like an actor playing different roles in different plays.
4. **Concentration**: Equally important. One slip of a tongue can flop a play. So does a wrong step in a math problem. The former is not reversible, though.
5. **Sequential nature**: Lines of a script are to be delivered in a particular sequence like mathematical steps. Deductive logic controls mathematical steps. In math, the tone has little to do with it. In a play, it can make a world of difference how a phrase or word is spoken out.
6. A mathematical theorem enhances one's understanding of problems once it is proved. Scriptwriter has a message too. It may be simple, deep, and of any kind; social, political, religious etc. Just like a math theorem is measured by its impact, so is a play. We all know the power that Shakespeare's plays still hold all over the world. Likewise, the Simplex Method, based on mathematical theorems, is a billion-dollar industry.

7. Math, by and large, is solo by nature. That does not mean collaboration is not there in mathematical activities. Apart from soliloquies, a play is rarely solo. It is team oriented.
8. You make a mathematical error, and rarely one is embarrassed publicly for it. In a play, the mistakes are publicly scrutinized.
9. Often a mathematical problem can be solved in different ways. But different proofs or solutions have little difference in impact. But a play can have varied impact depending on the vision of the director. A script may also be changed to suit a director.
10. During a play, the sound engineers, light engineers, costume designers, and stagehands play significant roles in the production of a play. Mathematics just stands by a mathematician without any fanfare, often relying on authors' papers, pencil, calculator, or a PC.

While watching these plays, do not get engrossed in the plays. 'Pinch' yourself **that you are here to see the connections with mathematical thinking**. I am not at all interested in just reading your summaries of the plays. Partly, I shall be there to watch them myself.

Let us together discover how mathematics and mathematical thinking are playing a role. Finally, both Black Box and Paul Harris theaters are excellent from the point of view of their small sizes and settings. You would be so close to the sets and performers that you may feel, as if you are a part of the play. If you have not experienced a small theater before, it is worth it.

(October, 2002)

CALCULUS DEFINES CIVILIZATION

During an Honors Council meeting, the Dean mentioned the upcoming review of the Honors Core Curriculum. **Currently, Honors students, not majoring in any science or engineering area, must take two math courses.** While expressing his views on excluding calculus from the requirements, he asked for my comments. I said, "Calculus as taught in a 3-credit HON 141H (College Mathematics II) is different from a traditional 4-credit HON 181H (Calculus I) designed for students in science and engineering." In fact, calculus in HON 141H is more 'diluted' than in MATH 176 (Calculus for Business) for which MATH 124 (College Algebra) is a prerequisite.

After the meeting, my mind stayed on calculus as a paradigm. I believe that the study of calculus is as important for the total development of young minds, as the study of Shakespeare in literature, Socrates and Plato in philosophy and critical thinking, and Aristotle in political science and history. Calculus sits on the concept of **Limit** that crystallized over a period of two thousand years. **In the entire gamut of human thought, there is no concept as profound, and yet as practical, as that of Limit in Calculus.**

There are misconceptions about Calculus both in its teaching and learning. Calculus in US high schools is considered a sign of prestige, though most students learn only how to spell the word 'calculus'. In colleges, math professors strongly believe (I, too, was one of them!) that the only way to teach calculus is through mathematical rigor, popularly called ε-δ – an approach only suitable for math majors. **The time has come to tailor calculus to meet the specific needs of different groups both in its concepts and applications.** I wish I had the discipline to have written the first serial book, *Calculus for Dummies!*

Calculus is truly a defining moment in the history of Western Civilization. Surveying contemporary civilizations since the 11th century, Europe was in the dark ages and America undiscovered.

Islamic empires were raging in the Middle East, North Africa and Eastern Europe. The Mongols ruling over China and Europe formed the greatest empire in history. Travelogues like that of Marco Polo describe the prosperity and glitter of Chinese royal courts, inspired generations of Europeans to seek their fortunes in Indo-China. The Mughals in the 16th century north India built one of the wealthiest empires. But mathematics and science did not make the considerable headway despite the knowledge of Hindu Numerals and its decimal system. **However, their power took off once the Hindu Numerals replaced the Roman Numerals in Italy in the 14th century.**

The discovery of calculus was a quantum leap out of the union of deductive reasoning of mathematics and experimental demonstrability of science. It actually burst into the firmament when Newton came upon the scene in the 17th century. He remains the greatest mathematician and the greatest physicist. It was at this time that the European Civilization caught up with others. **Calculus is at the heart of progress in sciences, and it became an engine of industrialization in Europe.**

Since the rest of the world had no clue of calculus, it continued to lag behind Europe in mathematical discoveries and scientific inventions. During the 18th and 19th centuries, it naturally propelled Western Europe to explore the world that also resulted in the colonization of Asia, Africa, America, Middle East, and Far East. Though by the middle of 20th century the colonizing had ended, the West continued to forge ahead. **The power unleashed by calculus in 17th century transformed into nuclear power in the 20th century!**

The magic of calculus lies in the fact that it lets one have the glimpse of infinity, and what is potentially infinite. Limiting process is inherent in human experiences. The common classroom scenarios helps one understand it; how, in a 'limiting' process, an average velocity becomes instantaneous, and how a line joining two points on a curve turns into a tangent at a point. Many ancient paradoxes like that of **Zeno** are laid to rest once the concept of **Limit** is grasped. Dividing by zero is a no-no in schools, but **I often**

characterize calculus as a story of Division by Zero, or story of Infinity!

Calculus is the ultimate experience of a rational thinking mind. Above all, **a grasp of infinity through calculus brings one closer to an understanding of Infinite Manifestation;** call it a god, or the God! So who would like to deprive oneself or others of its taste and beauty? As educators, we owe every student some exposure to calculus: a giant step of mankind in its intellectual evolution.

(June 09, 2003/Rev: Mar, 2008)

A Forty-Eight Hour Piece

A good story is worth telling again. It happened three weeks ago on Friday, the last day of Summer Session II. I took my car for an oil change and check-up on the brakes, and came back $400 short. I said the mechanic had lived up to the reputation of his profession. Despite my instructions for saving the old parts, they were not given back, thus leaving a bad taste. I told my wife how I felt ripped off by at least $200.

On returning home, I finished grading the final exams. Recognizing the top student in every course is my style of teaching. A girl Tina Arez won this distinction in one course. Next day, on Saturday, I called her to give this news. Her parents alarmingly picked up the phone. From their English, it seemed they were recent immigrants. They calmed down when I told them how good their daughter was in a mathematics course. I suggested them to encourage her to take calculus in the next summer session starting on the following Monday. They just listened, and did not say a word.

Tina, not being at home then, excitedly called me after a while. But she told me that she could not take the next course. Despite my advice that math courses should be taken without any gap, she just kept saying, "I can't do it." When I insisted on knowing one good reason, she responded, "**I have to help my parents by cleaning the houses before the fall semester begins**." It made me speechless.

I discussed this situation at home. Next day, on Sunday, I approached my daughter's fiancé, Alex whose heritage is Mexican. After explaining the situation, he spoke with Tina's parents on my behalf, and informed them that Tina's fees and textbook will be taken care by her professor. Alex, speaking in Spanish, convinced the parents that Tina's going to the college shall help them all in a big way, in near future.

It was Sunday afternoon; a few hours were left for the registration before classes to begin on Monday morning. On the phone, I gave

Tina my MasterCard number for charging the fees and register online. The amount was $350, but she only charged $250. It showed another side of her character, and convincing me that one day she would be successful in life.

In the afternoon, she and her younger sister came up to my office to pickup the textbook. She brought a pen, candies and a thank you card signed by the entire family. It touched my core. I reflected how in a span of just 48 hours the pleasure that I have accidentally derived from helping a brilliant needy student has far outweighed the bitterness over a rip-off by a mechanic. **I realized it as a different kind of investment; one human being at a time, as a phrase goes by.**

(Aug 04, 2003)

BRIDGING MATHEMATICS AND MUSIC

Mathematics and music are usually placed at the opposite ends of a spectrum of human endeavors. Music evokes a sense of natural universality about it. In contrast, mathematics appears to be a highly structured and cultivated discipline. From the layman's point of view, neither mathematics is strictly musical, nor is music obviously mathematical.

To the sensitive ears, there is music in the chirping of the birds, rustling of the leaves, and breaking of the waves. Great symphonies have been inspired by such settings. These places may be conducive for mathematicians to get away for a break when stuck with a problem. But it is naivete to imagine that one would get new mathematical ideas from such scenic spots.

Perhaps, the question of correlation between mathematics and music is more manageable, if one focuses on music generated by man (used in a generic sense). What kind of bridge is there, or can be built between mathematics and music? The underlying reason for this examination is that after all, man is the ultimate common entity. All forms of knowledge emanate from man. Hence at some 'ground zero' level of human consciousness, there has to be a meeting place for the symbols of mathematics and music in their most nascent stages. The objective is to get the best approximation of that manifestation.

Examining it by posing a direct question: does the knowledge of mathematics help in the mastery of music, or vice versa? Based upon my personal experience of 45+ years in mathematics, but only a couple of years in music, the finding is that one is not a prerequisite for the other. Nevertheless, one notices a preponderant presence of numbers and sequences in every facet of music. In vocal singing, the modulations are described by various intervals over which one holds and takes certain pitches. There are fractions of all sorts. Keynotes of various instruments have precise frequencies of vibrations. Natural numbers and fractions are all over in music in the simplest to the most sophisticated forms. Theory of music is mathematical, but not conversely.

Plato's famous thesis, ***Music of the Spheres***, in my opinion, is more than music created by the motion of planets in space. Plato school believed that the natural numbers describe all the physical laws of the universe. It is, indeed, no less mystifying that Plato who had such a strong belief in mathematics called it the music of the spheres rather than mathematics of the spheres, a more comprehensible description. In the great Greek minds of the caliber of Plato, mathematics and music indistinguishably coalesced! That is why the Greeks studied them together.

In the mundane life, music is highly auditory when it comes to receiving it. Its creation comes out of mathematics that is subliminally audible. That alone gives the understanding of beats. Mathematics has no direct dependence on any one of the five senses. It is truly man-made and yet transcends them. Finally, the child prodigies are discovered in mathematics and music alone. Perhaps, a study over children may reveal a viable connection between mathematics and music.

(Started in 2002 and forgotten; 'finished', Oct 28, 2008)
(To be continued!?)

Student advising at any stage, from freshmen to doctoral, is not to mould a student, but rather help in actualizing his/her potential. First week of classes has all kinds of confusion and excitement. On Tuesday, around 1 PM, on hearing a knock, I opened the door. A young student hesitatingly said, "I am registered in your Advanced Mathematics Topics (MATH 489), from 1-2:15 PM." While getting him seated, I told that my fall schedule was now different. However, on checking the website I found him correct. My curiosity was aroused as to how this student registered himself in an advanced topics course without consulting with the instructor.

This student has an interesting background. He is only 16, and has already finished all three semesters of calculus and linear algebra! He seems to be a product of some home schooling and community college. I don't think he ever went to a high school. Anyway, it pleased me to notice that such a brilliant student was attending UNLV. After discussing some courses, I suggested him to drop this topics course and add Differential Equations I (MATH 427). He was already enrolled in Discrete Mathematics (MATH 251).

A week earlier, another student had seen me. Three years ago, this student started UNLV with Calculus III (MATH 283) in the first semester. During the second semester, he was advised to take Elementary Complex Analysis (MATH 459) without a prerequisite course, Linear Algebra (MATH 330). With the result, he got a low B grade. To make it worse, he was approved to take a graduate course, Complex Function Theory I (MAT 709) in fall, during his third semester. He got a high C grade.

University regulations do allow undergraduate students to take 700-level graduate courses, but only after he/she has taken all the required courses and is mathematically matured. It is not a routine practice. This student repeated MATH 459 and earned an A grade, and now wants to repeat MAT 709 to remove a solitary C in his transcript while learning the material not grasped earlier.

As a special legislative audit is upon UNLV, assessment in the air, and Math PhD program on the anvil, student advising needs to be objective and individualized. Nurturing students alone will bring credibility to graduate faculty and programs. At the graduate level, rotation of courses and right mentoring can go a long way.

(Aug 31, 2003)

PROGRESS REPORTS

Instructors rarely forget their outstanding students. Though, reflecting the city of Las Vegas, there is a lot of transiency at UNLV. Still I like to keep a tab on the progress of my top students. It happened today when I ran into David in a parking lot. In last Summer Session II, being the best student in **Precalculus I** (MATH 126), I recommended him to register in *Precalculus II* (MATH 127) in Session III. Math courses are optimized when taken one after the other. David also sought my advice about taking *Finite Mathematics* (MATH 132) along with MATH 127. I told him, if he maintains the same study habits, then he could finish both. Well, David made solid A's in both courses. He is now enrolled in Calculus *I* (MATH 181) this fall. I am happy at his continued success and see my advice working. Usually, I don't recommend 6 math credits in a summer session.

Coincidentally, on the same day, I ran into Cristina who took *Precalculus II* (MATH 127) in Summer Session II. She did not want to take *Calculus I* (MATH 181) in Session III for some family reasons. Being one of the best students in the last couple of years, I could not see her mathematical talent go rusty. It took some efforts to get her registered in Calculus *I*. It pleased me to learn that she was one of the three students to earn A grades in a class of over 30 students.

Cristina's story does not end here. I could not believe when she told me that she was now registered in *Honors Calculus I* (HON 181). It is exactly MATH 181, but only open to the students in the Honors College. They same textbook has been used to cover the same syllabus. She was advised by the Honors College to take a certain number of courses with HON prefixes in order to get a degree from the Honors College! It doubly disappointed me, since it was on my advice and recommendations that she had applied for admission in the Honors College. Cristina's parents, being poor and illiterate, have no idea of college education.

With all the advising centers growing at every level, the advising is getting more and more impersonal. In Cristina's case it may damage her self-confidence. It is certainly wasteful of her time and money. Irony of advising is that most students argue about taking courses for which they hardly have any prerequisites. Cristina, at the other extreme, is willing to repeat the same material even after getting an A in it!

(Sep 04, 2003)

A RECIPE OF SUCCESS

At times, a story line is so good that you don't want to see it ended. Last Wednesday evening, on unlocking the office, I noticed a big folded manila envelope lying under the door. There was no name or address of the sender. For a moment, some 'terrorist' ploy flashed my mind. Nervously, as I started opening, it showed a bunch of dollar bills in it! In that state of wonder and confusion I did not notice a small yellowish slip stuck inside the envelope. It read: "Professor Bhatnagar, Thanks a trillion for every thing, Your Apprentice, Kristina -- ."

Kristina was the best student in my *Precalculus II* (MATH 127) section that I taught in summer Session II. Her performance was the best by a student during the last 3-4 years. Since math courses are best when taken without any break, I strongly suggested her to sign up for a calculus course. However, she refused to register in *Calculus I* (MATH 181) in Session III. It was discovered that the reason was the lack of family funds.

With the help of a Hispanic friend, we convinced her parents to let her take this course rather than assist them in cleaning houses in the remaining summer time. Her parents are recent immigrants and have little idea of the benefits of college education. It was Sunday before Session III classes were to start on the following Monday. I gave her my MasterCard account number for charging the course fees, $340. But she charged only $240. It was this amount in twenty dollar bills that she returned it!

Some times poverty does take a person's character away. Kristina is an exception. Charging only the amount the family was short of, and returning it on her own accord, speaks high of her character and family upbringing. The odds for her success in life are very high.

(Sep 05, 2003)

EPILOGUE

Kristina graduated from UNLV in 2007 with major in mathematics. She won the Bhatnagar Award for being the top major of the year! In addition, she fulfilled the Honors College requirements. Her attending college inspired her younger sister to study hard in high school, and she eventually went on to attend college on a scholarship.

(Oct 27, 2008)

Time To Recognise Retention Rates

"UNLV dealt low ranking in the US News college survey report" is a headline on Page 3 of the 09/03 issue of *The Rebel Yell*. What disappointed me was UNLV's rating (2.4 out of 5) slipping from last year. Amongst the 15 factors that go into an institutional evaluation, freshmen retention rate and graduation rate hit close to my current thinking.

Freshmen retention rate boils down to the retention of students mostly in 100-level courses. Faculty resists whenever it comes to introducing retention/drop-out rates as a measure of effective teaching. There are some interesting offshoots. Some faculty members maintain that it is their higher teaching standards that students, not meeting them, drop out of their courses.

Years ago, I questioned an administrator, how come enrollment at the end of three weeks is taken into consideration, and not as of Day 1? I did not get a clear answer. In the name of quality instruction, a tough quiz, or a snide remark in the very first week would discourage students to drop that instructor. If students have the prerequisites, then they must be encouraged to complete the course. This is my professional ethic.

Most students are known to repeat a course like, College Algebra (MATH 124) up to four times! It is a bad PR for mathematics too. It amounts to telling the students to drop their money before dropping out of the courses. Over the years, in some 100-level courses, hardly 15 % end up passing the course with a grade of C or higher. The rest withdraw, or get C- or lower grade. Consequently, they have to repeat the course.

A few days ago, a concerned graduate assistant told me how nearly 20 % of his students did not come in the second week. If math sections are going to expand to accommodate more than 40 students, then drop percentage is also going to increase. No matter who teaches the freshmen, it affects university's rating.

However, the time has come to implement retention rates from Day 1 to Day N. Software technology is already in place. It is easy to use a benchmark percentage in each course averaged over the last five semesters. Faculty must support it. It is in their interest. Usually a carrot approach works.

Administration should push it, if UNLV is to become a comprehensive research university by 2010. During a full day retreat on Aug 21, all universities issues were discussed to raise UNLV's Carnegie ranking. Nearly 150 participants included faculty, administrators, regents, alumni, students, professional, and classified staff. The President was the most active!

(Sep 08, 2003)

STUDENT ADVISING, OR ODDS MAKING

Yesterday was the first day of fall classes, and my students fill out an information sheet. One piece of data is the number of credits taken and the employment status. For example, in my *Finite Mathematics* (MATH 132) Section, only 3 full time students, out of 35, are not working. Also, there are 12 students who are taking at least 12 credits and working full-time. Twelve undergraduate credits qualify one for a full-time status.

On Day One, I do inform the students how much time they need to study in order to pass the course. While applauding them for stretching their intellectual and physical limits, I temper it with a warning based on the data collected over years. To make a lasting impression, I told them a recent story, typical of UNLV students.

On the first day of 2002 Summer Session II, on recognizing a student walking into my Precalculus II (MATH 127) class, I said, "Aren't you Tyon Hull, and you were in my Precalculus I (MATH 126) course two years ago?" With a surprise look on his face, he asked, "How could you remember me?" "How could I forget you?" I responded, "You had registered for four courses (12 credits) in a 5-week session!"

When I cautioned him about the overload, he did not take my free advice seriously. On the following Monday, I congratulated him for making it through the first week. However, he volunteered, that he had already dropped one course. I said even three courses were a lot. He dropped another course during the second week. When he did not show up in the third week, I knew he was gone out of my course too.

In a span of last two years, Tyon managed to pull a C grade in MATH 126. Still, he desires to major in Civil Engineering! When he told me about starting with three courses, I said, "The odds are not in your favor." He said, "I am 36, and have to finish this degree soon." I felt like telling him that at this rate the degree is going to finish you first! He has a high school going son. He dressed expensive and spent $20/hour for private tutoring, yet lasted only three weeks again this time.

My focus remains on advising. Should advising be mandatory? I think not, with student population like that at UNLV. Yes, it should be made available on web, online, FAQ flyers, advising centers, whenever a student seeks out. **To some extent, teaching and advising are inseparable.** As instructors, we advise on Day Number 1 and on Day Number N, as to what to do after a particular math course is over.

Above all, student advising is a microcosm of advising from parental to professional. When there is too much of it is freely going, then it has diminishing returns. A person ultimately learns from his/her own mistakes. That is the American way. You become a hero, if you beat the odds; stupid if the odds terribly beat you.

(Aug 27, 2003)

CELEBRATING ONE SUCCESS AT A TIME

It is one thing to say a few words of appreciation for some one, but it is quite awesome to have them engraved on a plaque. I am thankful to the Department and Dr Phanord, as its Head. For a writer's sensitive mind, it triggered a sequence of events. When was the last time in Mathematics Department, when we celebrated each other's achievements?

Last Jan, Dr Verma was recognized for his service to the Department for 35 years including 22 years as Department Chair. All three tenured and promoted colleagues were presented plaques of achievements. It was done for the first time. Phanord always calls it a celebration of success. His welcoming remarks for two new faculty members struck a cord of new collegiality.

Yesterday, he personally handed over a booklet to everyone on the Department's achievements of the last fiscal year, and its vision. It is quite a document and blue print. There is no denying that it has been a banner year in the history of the Department. After three failed searches for an external chair, three law suits, the Department split in two Divisions, put under administrative receivership, the Administration decided to hire a Head.

I went on the record to say that the Department needs a 'plumber' and a leader, not a research professor that the Department was not short of. With likely success in Mathematics, the Administration is going to hire more Heads in troubled departments. To the best of my knowledge, Phanord has a unique distinction of being the first Head in the Nevada System of Higher Education.

I would like to append one more scenario to the booklet. As soon as Phanord's hiring became a certainty, one staff member resigned, the other threatened to resign, and the third one was under some 'firing' line. The fourth staff position was taken away from the Department. Imagine Phanord joining the Department under one of such 'welcome' signs. His handling of the office situation convinced me then that the future of the Department is going to be better than it has been during the last ten years.

Thank you Phanord, and thank you my colleagues for this recognition. Have a nice weekend!

(Sep 12, 2003)

TEACHING CALCULUS FOR 150 MINUTES

I just finished giving final grades in **Calculus III** (MATH 283). The class started with 20 students, but 15 finished it. However, it met for 150 minutes a day; four days a week in a five-week summer session. In fact, there was a parallel section at the same time. That shows there is a healthy demand for required courses after 4 PM. Except for the first day, from instructor's perspective, the time flew by in every meeting. There were no designated breaks. Only 2-3 students used to take short breaks. My only break was when the students were taking a quiz.

Teaching any math course for 150 minutes is challenging. It reminds me of the late comedian, Bob Hope's famous line on McDonalds, when the hamburger sales touched first one billion. McDonalds signs read – 'one billion hamburgers served'. During one of his NBC Specials, Bob Hope quipped, "How many eaten?" An instructor's teaching for 150 minutes does not mean that students equally have received it well. It constantly concerned me.

Three things were devised to keep the students actively engaged. Asking one or two students to come and do a problem on the board. During the first two weeks, five team quizzes were given to promote small study groups outside the class. Then we recognized the alert student (s) of the day; the ones who would catch my error out of my fatigue or oversight, or ask sharp questions during the lectures. Still, measuring its effectiveness from students' perspective is an open question.

Last time, I taught this course was solid five years ago. All the gaps of five years are not the same. Only the time can tell the difference. It was interesting to note that out of 20 students, 14 were taking it at least for the second time. This kind of data may be of special interest to new faculty members.

UNLV students are unique in the nation when it comes to work and study. Studying is not work here, as I knew back in India of

1950's. Work is not studying either. Most students work full time during summer. No one failed the course, but only one earned an A grade. One student took Calculus III and basic Computer Science course (CSC 136) in one five-week session! I could notice his health declining in three weeks.

In sports, Americans always support the underdogs. **I tell my students, if you beat the odds, then you are a hero. But if the odds beat you badly, then you are stupid!**

(Aug 17, 2003)

You Never Forget Your First---

Yesterday, a few students came up to me after the class to pick some handouts that they missed on Day Number One, and also to sort out their registration woes. Suddenly, one of them said, "I am Amanda, and years ago my father Ken Guralneak was your student, if you remember him?" "Sure I do," I said, "Ken was a graduate assistant and took my differential equations course. Soon after graduation, he joined Nevada Test Site. Later on, while working at EGG he also taught part time for the Department." In a way, I knew more than his name about her father. Amanda smiled and the image of 6'+ tall Ken flashed my mental screen as tall Amanda gracefully walked away.

Now here is a line. Amanda must have been impressed on the first day of class that on returning home, she talked about it with her father, who in turn still remembered me! I must have made some impact on Ken, though I don't recall any thing specific. Realizing Time is an ultimate eraser; it really made my day!

A similar incident happened in the first week of last spring. My memory was triggered on noticing the last name Hollingshead in the roster. I asked the student if his father was ever a UNLV student? He said yes. I joined UNLV in 1974; Guralneak was my student in fall semester and Hollingshead in spring 1975. That is a coincidence!

I do not possess a good memory, but you never forget some of your first experiences in any walk of life. In no other profession one gets so many opportunities to profoundly affect the lives of the individuals as in college teaching. The mystery of the whole process is that many a time it takes place without any one being aware of it! Nevertheless, happiness comes in small packages.

(Aug 29, 2003)

CAN YOU SHOWCASE MATHEMATICS?

I was out on the campus mall stretching my horizons and breathing fresh air. While passing by Lily Fong Geoscience Building I detoured into its first floor to see their showcases containing exotic looking rocks and minerals. The collection has been growing for the last many years through donations and faculty collections during field trips.

One showcase contained some fossils with well-written short notes on how they are created in nature. It flashed my mind that there are all kinds of artificial diamonds today, but no artificial fossils yet! For a few minutes, I stood enthralled by their beauty and awed by a million-year complex process that Mother Nature takes to create a fossil. It was not the first time that I had seen a fossil, but a new thinking it generated. Suddenly, an old timer interrupted my little 'study', and we walked out of the building.

While returning to the office my thoughts converged on how mathematics can be showcased? Mathematics has great and popular theorems like Pythagorean and Fermat's Last Theorem, but can they be displayed for casual visitors walking through the Dept hallways, or coming to Math Office? It is challenging to find their public appeal. Perhaps some latest applications of mathematics can be put together.

Yes, there are art objects driven by mathematics. During math meetings and conferences, arts and sports are the latest areas of mathematics applications. The Dept can buy math inspired art pieces and put them in one showcase with mathematical notes of explanation. The point is that in a media driven world today, if mathematicians won't showcase their products, then who will ever know and do it for them? A business thrives only when lot of people walk into its premises. Every business entices its customers.

On walking back to the office in the hallway, I gave a fresh glance to the notice board out side the Dept. workroom. There was really nothing of public interest. Besides flyers on job openings and graduate assistantship announcements, there was a big poster on a conference on Banach Algebra. That is OK, but has little PR value. People now use web sites to get such information. In a competitive world, it is worth showcasing individual mathematical activities whether they pertain to research, teaching, or service for some public consumption.

(July 03, 2003)

WHEN AND WHO TO TEACH?

In graduate schools, there is a time bound on the course work done towards a PhD degree. I remember Indiana University had seven years after all the requirements except thesis were done. At UNLV, it is 6-8 years depending whether one is admitted into a PhD program with or without a master's degree. Essentially, it means that if all requirements are not finished during such a period, then the course work is reset to zero. I personally know more than one individual who drifted away from their academic programs, and when they returned to graduate schools, they were told to start the course work all over again.

What makes me wonder is that professors seldom apply this principle on themselves. They strongly believe that since they have earned a PhD they are competent to teach any graduate course. I have known a colleague giving an independent study course in an area that he never taught as a regular course. He himself took it 20 years earlier!

A few years ago, another colleague remarked that he could teach any course. I said, "You are then equating teaching with spitting!" Unless an instructor is at least 3-4 courses ahead of the students, or/ and is actively involved in research related with the area that he/she intends to teach, it is unprofessional to teach that graduate course. It undermines what teaching is all about.

Spreading one's teaching expertise at a lower and some upper division courses is a different ballgame. I am talking about solid 700 level courses. Personally, 1 recall my experience of teaching *Real Analysis*, MAT 707-708, using Royden's classic textbook, during 1978-79. I took this sequence at Indiana during 1969-70 followed by functional analysis (from Halmos) in 1970-71. However, for PhD, real analysis was neither my major, nor any of the two required minors. Actually, teaching of this sequence more or less fell into my lap. I had a very tough time including some embarrassing moments!

Often, teaching some graduate courses is more challenging than some research activities. As the Department heads toward a PhD program, it is time that the faculty exercises some quality control on it. Integrity of graduate teaching is at stake.

(Aug 03, 2003)

EXAMINING GRADUATION RATES

Graduation rate is one of the dozen benchmarks in the institutional evaluation. UNLV's ranking slipped this year because of it, as announced in the **US News and World Report.** I know the strides UNLV has made, since I joined in 1974; it is just phenomenon. The drive of the administration is to propel UNLV at a higher level in a short time.

Graduation rate has two parts. One is the number of students graduating per year from a program at different levels; from bachelor's to doctorate. In mathematics, the figures have hardly changed over the years. For example, the number of undergraduates getting BA/BS in math has been constant, under 10 per year. It is amazing when one compares the number of students UNLV had 30 years ago to what it has today. Without strong incentives, the undergraduate data is unlikely to change in the next few years.

It is unaffected by foreign students who mainly come for graduate degrees. At the graduate level, I once tracked down all the students who earned MS degrees in math during a 10-year span, 1980-89. This number is exactly is 15. The number during 1970-79 may be around 18-20, as during this period, the EPA, housed on UNLV campus, encouraged its employees to do graduate work in statistics. The number during 1990-99 is close to 20. With a PhD program on the horizon and addition of a new concentration in Teaching Mathematics in MS degree, there is likely to be a quantum leap in the number of MS degrees in the next decade.

The other part of graduation rate is the number of years it takes to finish a degree. The total time period can be reduced, if students are given the option of finishing MS based upon course work alone. The current options are thesis or written exam. At Indiana University, I got my master's just on the completion of the course work.

To the best of my knowledge, only one student has chosen to take the written exam option since it was introduced 15 years ago. Students tend to drift at the time of thesis for reasons of their inability to tackle a thesis problem, or not getting right guidance from thesis supervisor, or both. It does require extra discipline on the part of the students. Mentoring graduate students factors in personalities on one-to-one basis. Mentoring plays into the dynamics workload and merit awards of faculty.

UNLV's Mathematics Department was already well established in 1960's when a new program in Hotel Administration was started. It was the vision of the first Chair/Dean, Jerry Vallen for 20 years, who hooked the program with famed hotel industry of Las Vegas. Today, **UNLV's Hotel Administration Program is Number One in the country**! During the next ten years, mathematics can take off, if its curricula and students are tied with other disciplines and local industry.

(Sep 28, 2003)

A PERSPECTIVE ON STUDENT CREDITS

An interesting dialog took place yesterday. Institutional data reveal that every year Mathematics Department generates more student credits in the College than the rest of the four departments combined. Translating it into dollars, it runs into a staggering figure of over $2 million for 2002-03. Yet, mathematics is the only department without its own building! If students to faculty budgeted ratio of 15:1 is taken, then Math Department is also under staffed despite including the part time instructors. How do you explain this anomalous situation in the university existing for the last forty years?

Such a scenario can not exist in athletics. A sport that brings money gets a lion's share before it is distributed to revenue losing sports. Most of us would recall, how in 1980's, the UNLV basketball team was the toast of the town. The Running Rebels playing in Thomas and Mack were another show in town competing with Strip. It was also a place to see and be seen. In 1973, the Rebels played in a 6,000-seat hall of the Las Vegas Convention Center, but in 1983, they moved into 18,000+ Thomas and Mack basketball arena. Through 1991, it was always sold out. The Rebel basketball put UNLV on a world map. The money flowed into UNLV coffers like water gushes into flood channels in Las Vegas valley after a short cloud burst.

The question is why such a disparity in the treatment of two programs; one in academics and the other in athletics? Fans of Running Rebels are not a 'captive' crowd. They simply follow the winners. Most students in lower division math courses are 'captive' and they hate taking these courses. At least 50 % of students either drop the courses or get a grade less than a C. Consequently, either they change majors, or repeat the course. **The entertainment value of mathematics is relatively minimum**. At the upper division level, students may not hate math, but they are not rushing to become math majors. Out of some four thousands that graduate each year, only a total of 6-8 major in mathematics.

The ratio of the number of students that graduate in a program to the number of faculty is a good measure of a department's productivity. I also think the data on student credit should be analyzed in all three categories; Lower Division courses (100-200), Upper Division (300-400) and Graduate (600-700). By and large, students in upper and graduate level take courses by choice.

Students, particularly from out of state and foreign countries, come for the reputation of a program whether for playing a sport, or doing a PhD. Unless an academic program has some stature in terms of its infrastructure, quality of students and faculty, any argument for new resources based on student credits alone won't carry it very far.

(Oct 02, 2003)

CULTIVATION: MATHEMATICS VS. ART

Have you ever heard of a person becoming a 'reasonable' mathematician without going through some formal college education? At least, I have not. In the world of art and music, men and women have achieved proficiency and fame by dint of their sheer individual efforts at homes and studios.

Currently, UNLV Barrick Museum is hosting a unique exhibition of paintings. All the artists are self taught. They did not attend any art school, and came from humble beginnings. Some prodigies started painting at a very young age. They are nearly fifty of them from all over the US. Noticeable similarities are seen in their brush strokes, choice of colors, finish, and a lack of sophistication. There is no limit to the use of material.

Some artists got into painting during or after a long illness, crippling tragedy, or from a visionary epiphany. Their work, by and large, is grounded in solid imagery. It does not fall into any school of art abstraction. In mathematics, algebra is a quantum leap into a mathematical abstraction. Geometry is intuitive, but its axiomatization is abstract. Uncultivated persons in mathematics are least likely to get any public recognition. In contrast, the art galleries and museums are going after the works of self-taught artists for their boldness and uniqueness.

It reminded me of a man in his 40's, who, in 1970's, nervously walked into Math Dept holding a brief case full of work sheets close to his chest. He was recently released from a prison. His formal education was hardly at middle school level. However, in prison some how natural numbers caught his fancy and he spent days and nights playing with them. He claimed to have discovered some results that would make him a millionaire! However, he did not know even formal intermediate algebra.

The Dept offered to help him polish his natural skills and learn some college math, but he did not want to stay. **It may have been**

no different event from what Hardy and Littlewood discovered in Ramanujan 100 years ago. Mathematicians are cultured by formal training and mentoring.

I told my students that no matter how much one is gifted in math, he/she is like an uncut diamond. Cutting and polishing alone make it valuable. It is only with the understanding of **deductive reasoning that one gets a grasp of the concept of proof in mathematics that has not been known in many ancient civilizations.** That sets art apart from mathematics.

(Oct 09, 2003)

OBSERVING ART AND MATHEMATICS

What is common between Art and Mathematics? This question popped up in my mind as I gave a first glance at the paintings on all four walls of an art gallery. They are the works of Mary Warner, a UNLV Art Professor. An answer immediately surfaced up: **Paper and pencil!** Paper and pencil usage has dominated mathematics till the end of the WW II. There was only one kind of mathematics, and even theoretical physics was a part of it. In 1993, with paper and pencil alone, Andrew Wiles proved a 350-year old unsolved problem, *Fermat's Last Theorem*. So did Einstein in discovering great theories of relativity in physics, though its predictions were verified much later.

I also do all my mathematics with paper and pencil. Paper being scarce in India until the 1950's, we used slates and chalks to do scratch work. Paper and ink were meant for fair work. **That is the secret of Ramanujan's famous Notebooks!** The *Beginning Art* course (ART 101) is taught with soft lead pencils and drawing paper. Several artists do pencil sketches including Picasso, who did a dozen of them. Some photographers believe that black and white photography brings a 'depth' into the pictures. Certainly, no one produces a black and white movie any more. Actually, the old ones are being dubbed into colors.

There is something striking about these paintings, though they are all of flowers in ones, twos or threes. What stands out is the artist's rendering of mathematical concepts of limit, convergence and infinity. **That makes these notions innate and intuitive**. You intently look at any painting, then a point is there to which the petals and shoots are converging to, or diverging from, depending on your focus on the flow. **There is a perception of movement in her work**. She has chosen flowers that exhibit a pattern of 'continuity' over 'discreteness'. Thus the concepts of limit and continuity do not belong to the exclusive domain of mathematics. Mathematics only gives unambiguous meanings to these terms, whereas the dictionary meanings are varied.

The paintings are in bold, pastel, and soft colors. They are done in various sizes and medium. A couple of them are 18" x 18" and one is 48" x 72". **They may be like math research papers from a smallest publishable unit to a complete monograph that completely settles a problem**. It seems that quality and quantity in research factor into art academe too! Perhaps, it is universal.

Another feature of her work is the use of different surfaces: flat, convex and concave. Curvature heightens spots of darkness and brightness. Sure one can write mathematics on any surface, but it serves little purpose except adding weirdness to it.

(Oct 21, 2003)

PS: If you are curious about the exhibit called *Avant Garden*, then it runs in the Donna Beam Fine Art Gallery through Nov 8.

UNIVERSITIES AND PART-TIME FACULTY

UNLV prides itself in being called a metropolitan university. It is the only university situated in the fastest growing city in the country. Ever since my participation in a university planning retreat last August, off and on, I have been trying to identify some of its metropolitan characteristics.

In 1974 when I joined UNLV, Las Vegas was not on the map of the retired communities. Florida was the only destination. In the 1960's, Arizona became the second choice as water and power brought greenery and central air conditioning to the desert. It was in the 1980's that Las Vegas came up on the map, and has now surpassed Florida and Arizona in attracting senior citizens from all over the country. Seniors bring talents and economic boost to a state.

I still recall Leo Schumann, a retired engineer who 'got sick of puttering around the house' and became one of the best math graduate students of his class (1976) at 68. He taught part time for the Department for another 10 years. Leo's contemporary, Art Bell, at 80, took graduate math courses in order to 'keep his mind sharp'.

UNLV, realizing a potential in senior resources, started a Senior Citizen Program under which Nevada Seniors over 62 pay no tuition fees during the academic year and 50% during summer. As the senior population grew so did their academic interests. **EXCELL, Elder Hostel** and **Senior Adult Theater** programs, offered through the Division of Extended Education, are so successful that they have brought national recognition to UNLV.

As part-time instructors, senior citizens can make significant contributions to UNLV. The Dance Department, amongst its adjunct faculty, has a large number of seniors. **However, UNLV needs to tap this resource systematically.** I have encountered retirees with graduate degrees including PhDs from top notch schools with distinguished professional careers. Most are financially well off, but they want their minds engaged in a challenging manners. The benefits of their association with UNLV are far reaching, if such contacts are properly tapped. They are potential donors too!

Look at McDonald's and Wal-Mart, the two most successful business houses in the world. Before 3 PM, most employees at McDonald's are senior citizens. In Wal-Mart, senior employees are seen working as greeters and helpers. **There are remedial and lower division courses in every department that are cost effective, profitable and quality controlled, if taught by part-time instructors.** Research/PhD faculties have no aptitude for these courses! "Mind is a terrible thing to waste" is a famous ad line of the United Negro Fund, applied to teens' potential wasted for lack of resources. However, it also applies to the communities that let the capabilities of their seniors go untapped.

(Nov 23, 2003)

A Look At Examples In Textbooks

"How many of you have worked out all the eight examples?" I asked last Tuesday. No one raised a hand! **I said to myself that at least my students are honest.** These examples were part of the home work assigned a week earlier before Thanksgiving Day. Besides these examples, there were ten standard problems in the context of the course, MATH 132 (**Finite Mathematics**).

In all lower division courses, worked examples form an important part of my instruction. On Day Number One, I openly tell that **on every test there will be a problem already worked out in the text, or discussed in the class.** Reasons are simple. Mathematics is very unique. You learn math only by doing it with your own hands, never by watching others do it. You hear a song you may remember all of it, but you hear math, you soon forget it! You may be able to understand a lot of literature, history and psychology by reading it in the bed. One can not grasp math by reading it; better to sit on the edge of a chair before working out the problems.

After each test, students are amazed to find how true this assertion is. In social sciences there is nothing like 100% understanding of a concept because of subjectivity involved in those disciplines. **But in math, there is nothing less than 100% that is right**! Most students in lower division courses are not math majors. However, this periodic exercise on the nature of math registers on their minds. They may become better math students in the long run.

About examples, I have a couple of other observations. First, **beyond 3-4 examples, there is a diminishing return in term of students' understanding.** A large number of examples overwhelm the students. If on the top of 8-10 examples, an instructor does 4-5 additional problems in a class, then it is too much of a 'spoon feeding'. **Even a baby turns his/her mouth away when too much is spoon-fed by the mother!**

Students no longer study all the examples in a text book, or problems discussed in the class before attempting homework problems. Still some students complain in course evaluations that there are not enough examples in the textbook or done in the class. They demand one-one to correspondence between the examples and the variety of exercises!

I recall the first edition of *Finite Mathematics* textbook by Mizrahi and Sullivan, the number of examples was less than 50% of the present 8th edition. The number of pages was two thirds of what it is today. **Students pay more for a book fattened with examples.** The textbook prices increase as number of pages increases. Sometimes, I also wonder at a correlation between the rise in body weight with the increasing weight of lower division math textbooks. **It is time to trim them down!**

(Dec 04, 2003)

WHAT IS ON THE CUTTING EDGE?

Yesterday, I was talking in the hallways about attending the world's largest gathering of mathematicians in Phoenix (Jan 6-10). "**What is the latest in mathematics that will be announced at this annual meeting?**" asked my Department Head, an experimental mathematical physicist. I did not have a good answer to this question. It is a joint meeting of six mathematics organizations: The American Mathematical Society (AMS), Mathematical Association of America (MAA), Association of Symbolic Logic (ASL), Association of Women in Mathematics (AWM), National Association of Mathematicians (NAM), and the Society of Industrial and Applied Mathematics (SIAM).

However, I continued to mull over this question. Having attended many such meetings, **I can not recall one meeting when some deep mathematics results were announced with fanfare.** Presidential addresses are generally survey types on the state of mathematics. Mathematics results rarely catch the attention of the media except one in 1976 when Rene' Thome announced his seminal work on **Catastrophe Theory** in a summer meeting. May be the word Catastrophe was too catchy to be ignored! But it did generate a varied interest in mathematics from applications in stock market to biology.

Annual conventions of every trade organization provide a forum for the release of their latest products. Physicists have headlined the properties of subatomic particles and unification of forces. The media always catches them. No wonder physics remained the darling of every granting agency from 1940's through 1980's. Likewise, astronomers love to report on new stars, and biologists on new bacteria and viruses. In Dec 2001, I happened to attend the conclusion of the annual meeting of the America Geophysics Union in San Francisco. **In the evening, the first shots of the core of the sun were shown to an invited group**. The clip, hardly lasting a few minutes, reportedly cost in millions, qualifying it as the most expensive video/film production in the world!

It is time to re-examine the program of the joint meetings. Despite the breadth of mathematics it is possible to identify two, if not one, deep mathematics results of a year. Likewise, some application of far reaching consequences must be identified. Achievements of women and minorities in mathematics easily draw the press. We are living in media savvy world. It is time to tell the world what mathematics can do. **If mathematicians won't brag about their discipline, then no one in the world will do it.**

Annual highlights can be used as great motivators at various levels of in-class instruction, research and grant activities. During the five-day conference there are invited lectures on assortment of topics, **but rarely one special lecture or press release is targeted for the public at large.** After having been in the profession for over 40 years, I strongly believe that it is possible to make a tangible connection between any abstract mathematical result and some aspect of the world around. **All it needs is a change in attitude.**

(Jan 06, 2004)

ON POPULATING GRADUATE COURSES

Yesterday, two diverse situations converged to a point. **Enrollment management of the College comes under the charge of the Associate Dean.** The Associate Provost sends out periodic reports to the colleges on registration updates. The departments are then constantly reminded to open up new sections of some courses, increase the enrollment caps, and cancel some sections. Therefore, it is essential to know the 'popular' (that draw students) courses and 'unpopular' (that do not draw students) courses in each department. I noticed that 700-level Topic courses in some departments were under a flag of monitoring or cancellation.

The Graduate College requires that a description of each Advanced Topic Course and Independent Studies course be on the file for record. **Ideally, instructors should post and publicize them on bulletin boards.** On the same day, an undergraduate student petitioned to the Dean's office for taking a 700-level *Advanced Topic Course*. Giving a hearing to such students is also a charge of the Associate Dean.

The student was quite upset despite the fact he was counseled by two faculty members and the Department Head. On examining his transcript in detail, it was discovered that the student had not taken even a single course 'required' to take this 700-level course. After an hour, a first year graduate student came over for taking a 700-level course. On checking his math background, it was found again, that this student had no prerequisites whatsoever!

A couple of such cases come up every semester. More faculty members want to teach graduate courses, but the problem of attracting students to math degree programs remains acute. It is not ethical to 'recruit' undergraduate students into graduate courses without a strong background. **The provision of undergraduates taking graduate courses is only for well prepared exceptional students.**

The Department has an excellent array of upper division 400-level courses for the undergraduates to bring breadth and depth in their math background. A 700-level course without any pre-requisite is a 'suspect' graduate course. It is time to exercise a quality control particularly when math PhD program is on the horizon.

(Jan 14, 2004)

THIS IS NOT A CATTLE FAIR

Attending the annual Joint Mathematics Meetings is always a great experience. It is a place to arm wrestle with ideas and do some sparring. You stand up to share your bit of innovation, and then listen to others. This year (Jan 5-10) it was a spectacle of the world's largest gathering of nearly 5000 mathematicians from every cadre assembled in Phoenix, Arizona. They come from every part of the world if they can get the US Visas and travel support. From the facial contours, I would venture to say that the number of attendees from Asian countries was smaller than before. It is due to greater visa scrutiny since 9/11 aftermath; America is at war.

Professional meeting is a place of networking, and to renew and make new connections including preliminary interviews and negotiations between job seekers and employers. A big attraction is the exhibition of mathematics books. Browsing the display, I sensed a hidden **Existence Theorem: Given a mathematical topic or theme, there exists a pair of author and publisher to bring a book out on it.** Such a scenario could be encouraging as well as discouraging to a new writer.

The major business is the presentation of nearly 1500 papers. Without exaggeration, it is plausible to say that the range of topics is potentially from minus infinity to plus infinity. The sessions are punctuated with scintillating lectures, catchy titles sponsored by one or two of the six math organizations that put this whole show together. **I did not have any thing to complain about the logistics of the convention in Phoenix, except why not to have one in Las Vegas too**? After one and only one 1972 Joint Meetings in Las Vegas, the Math Brass resolved not to hold it in Las Vegas. With the support from the administration and Convention Authority, this decision can be reversed. It will enhance the image of UNLV, and particularly that of its Mathematical Sciences Department.

Talking of the Meetings as a Math Show, it triggered my association with the Indian Mathematical Society and Malaysian Mathematical Society. I became their life member while I was in India and Malaysia for two years during 1980-82 and 1992-94 respectively. I never received any information on their annual meetings or journals as a part of my membership! **Reflecting, that is what sets USA apart from these countries**! Behind any well-organized event there is team of dedicated people and skilled volunteers. I have come to a conclusion that Americans are far more successful in organizing mammoth events than any other national group.

A funny thing about the conference was that it reminded me of great Las Vegas buffets! There is an interesting parallel. If you go to a dinner buffet, say, at the Rio Hotel and Casino, the two eyes cannot see nearly 200 entries (including toppings, dressings, salsas and salads) in one glance without turning them through 60 degrees. How many items does one ever eat? 10-15 depending upon the appetite? Well, nearly the same scenario is encountered during a math meeting. Even if you walk in and out after listening to one talk, you may not be able to attend more than 100 presentations. That will numb the brain! It is amusing to see people run from one presentation to the other during their 3-4 days of stay. They are bound to be constipated at the end! **Listening to new ideas is like weight lifting**. If you have not been doing it, then sudden lifting of the weights can damage and sprain some body parts.

I literally ran up to a one-hour AMS session, **Who wants to be a Mathematician?** expecting it to be a panel discussion of mathematicians on intellectual challenges and opportunities for math majors. A big hall was packed with people standing along the walls. However, it turned out to be a math game featuring 10 high school students competing for the $2000 Grand Prize. It was organized exactly on the format of the popular TV show-- **Who wants to be a Millionaire? Pursuing great ideas can become fun only after a lot of hard work and sweat.** Generally, what begins with mere fun ends in a fibster. That is how I felt about including such games into a course curriculum.

In the annual meetings, I usually search for newer applications and frontiers of mathematics. Interest in **Math on the Web** is on rise, as measured by the crowds in the room. On a 'lighter' side there was a panel discussion on the **diverse personal lives of mathematicians** and an evening reception. MAA special presentation **Cinemath: Mathematics of the Sliver Screen** got a little wind in its sails after the 2001 popular movie, **The Beautiful Mind.**

For math instructors interested in professional development activities, there were nearly 30 minicourses spread over 2-4 hour sessions. Topics ranged from mathematics of card games to some aspect of teaching. Since the last decade, mathematical biology is showing dominance. The prestigious Gibbs lecture was on **Biology as Information** by Eric Lander and an AMS/SIAM paper session on **Mathematical Modeling in Neuroscience, Biomedicine, Genetics and Epidemiology**, and a stirring AMS-MAA invited lecture by Bonnie Burger of MIT, on **Mathematical Challenges in Molecular Biology**. It is time that UNLV faculty in Math and Biology explore a viable joint bachelors and masters degree programs. Without a nursery, there can't be a plantation!

For the last couple of Meetings, there has been noticeable rise in the stock of **Mathematics and Arts**. There was an art gallery in one room, and display of sculptors and artifacts in the other. Some of the works, inspired by mathematics and mathematicians, were valued in thousands of dollars apiece! Apart from visual display, there were **three** paper sessions on **Mathematics and Arts**. In order to ignite interdisciplinary interest, I have sent sets of the materials to the Dean of the College of Fine Arts and Chair of the Arts Department.

The place where people did not even have a standing room in a 100 plus capacity room was in a session on **Mathematics and Sports**. Perhaps the NFL playoffs added some fuel to it. Research in this area has been on a steady rise. It is not all statistical, but good mathematical modeling in some cases. After all, one NFL game impacts the national economy in hundreds of million dollars!

There were **three** AMS-MAA paper sessions on **Mathematical Techniques in Musical Analysis** and a musical presentation on the **Mathematics of Acoustic Paradoxes** by Erich Neuwirth of

University of Vienna, Austria. A traditional college curriculum has set math and music far apart from each other. During the time of ancient Greeks, math and music were welded together. **Music of the Spheres** by Pythagoras speaks of their inherent unity. **The child prodigies are only known in the domains of math and music!**

It was 20 years ago when I presented a paper in an MAA session on **Teaching Mathematics through History**. Since then **History of Mathematics** has grown into a special interest group in MAA. **The number of papers and talks in various areas of History of Math far exceeded than on any other topic**! AMS and/or MAA sponsored several paper sessions. After all, mathematicians also pursue name and fame, and hence are drawn into its politics and financial issues. That is what the history of people is all about.

Finally, what stood out was the AMS Retiring Presidential Address by Hyman Bass, University of Michigan and one of the few persons who is a member of both National Academy of Sciences and National Academy of Arts. In his talk on **Mathematics, Mathematicians, and Mathematical Education**, he essentially said **two things**. One was explicit, that research mathematicians should take charge of all level of school math instruction. The second was implicit; go after big education grants even at K-5 level and engage math educators to work on those projects while you continue to do research in abstract topics and obscure areas of mathematics.

I disagree with the first, but agree with the second. New Math of 1960's, as pushed by university mathematics departments, miserably failed a few generations. In the present time of public accountability, mathematical research has to be related with applications.

My personal engagement in the Meetings was as an officer in a new special group of MAA, **Philosophy of Mathematics.** I also co-chaired the second session on it. At the end of every day, I unwinded with beer in an 18-oz can!

(Jan 20, 2004)

A Low Hit On My Teaching

Last week, while filling out the Y-2003 Annual Report, I was kind of shocked to notice that in *Honors Mathematics II* (HON 141H) the GPA (Great Point Average) on students' evaluation of my *Overall Instructor Evaluation* was 2.00 (**Excellent** has 4 points, **Commendable** 3, **Satisfactory** 2, **Poor** 1, and **Very Poor** 0). My first reaction was that had it been even 1.99, I was out of merit consideration for the entire year! The GPAs in *Discrete Mathematics* (MATH 251) and *Finite Mathematics* (MATH 132) were 3.00 and 3.52 respectively. Most students in these courses are non-math majors.

Such are the moments for my *Reflections*. Did I expect 3.0; the highest GPA an instructor has ever gotten in HON 141H? My answer is No. Teaching calculus to students, not majoring in any science and engineering areas, is very challenging. Honors students often complain against this requirement. At the end, they do take it out on the instructors.

Am I an effective instructor? I really believe it so. Despite having won the **College Distinguished Teaching Award** and runner up several times, I tell my students that I would earn this wing from you being a unique group. Complacency does creep in any human activity. Jimmy Johnson a winner of two Super Bowls with Dallas Cowboys could not turn Miami Dolphins around and thus resigned. Pat Riley a legendary coach in NBA failed with Miami Heat.

The psychology needed to motivate players or students is the same. However, motivational techniques can go out of style with new social norms. I felt it last spring. To turn the attitude around, extra credit assignments, bonus homeworks and other incentives were given. By and large, they all fell flat at the evaluation time!

Do I give a lot of high grades to receive high GPAs from my students? Never! In HON 141H, no one evaluated me **Excellent**, and B+ was the highest grade that only one student earned. However, one gave me an **Unsatisfactory,** but none failed this course. The lowest grade is a D with cumulative GPA of the class at 2.28.

I had informed the Honors College, not the students, that I would take a break from this math sequence. Unless one is writing a textbook, teaching the same course semester after semester, or even year after year, can lose its steam. **It is better to get away and have a fresh perspective when the student GPAs are low, or falling.**

Some students do 'gang' up while filing out the evaluation forms. This was a class of 16 students from nearly 16 different majors, and no one had dropped out! Honors students, usually taking the same courses in the first year, tend to share the teaching evaluation. Students are aware of the long term impact of their course evaluation on faculty merit, promotion and tenure. Nevertheless, I am, and have been, in favor of student evaluation of all the courses they take. That is the American way of life, in general.

(Feb 04, 2004)

A Paradox In Communication

Public speaking is the second biggest fear amongst people. The first, of course, is the fear of death. In order to develop and hone my speaking skills five years ago, I joined a toastmasters club. **Communication has two main parts, listening and speaking.**

In a club meeting, it takes only three seconds to walk up from a seat to the lectern. Often I forget my opening lines in **three seconds**. After returning to the seat, I then wonder how I could forget it! Though overall anxiety is getting less, but a 'sudden loss of memory' at the lectern does recur in different scenarios. Fortunately, it does not freeze me up.

I have **two** analyses to offer. One, in anticipation of my name being called on stage causes excitement and pounding of the heart. A rush of adrenaline may trigger a sudden 'rewiring' in brain cells. It may be like a surge of power voltage that trips an electrical circuit, and goes over to the other. That is why I usually don't end up speaking what I wanted, but speak out that not planned.

The other angle is that as soon as I am in front of the audience, the very process of their looking at me brings a sudden change in my neurons. **After all, any process of observation is a shower of photons!** It is a fact that cheering or booing from an audience affects the performance of players and entertainers, in particular. Sightless persons are not affected in the same measure.

As the President of TNT Toastmasters Club, the initial and final segments of the meeting fall into my duties. At the end, there are only 3-4 agenda items taking no more than seven minutes. Last Monday, I forgot to announce the winners of the day for the third time in a row! **That is what I call a Paradox in Public Speaking.**

(Feb 25, 2004)

Note: This write up is a part of the material that is distributed to the students in the Honors Seminar, *Paradoxes in Arts, Science and Mathematics* being taught this semester.

Challenge Of Remedial Courses

Last Monday, I told a colleague that even after 30 years of teaching at UNLV, I was feeling butterflies while walking to the first class of **Elementary Algebra** (MATH 095). As a matter of fact, I have avoided teaching this course! The last time I taught **Intermediate Algebra** (MATH 096) was in Spring '80! **When a course gets to that level, for a PhD instructor the entire challenge shifts to its effective delivery.**

For years, I believed that during semesters, it is underutilization of PhD faculty to teach courses below Calculus. For a similar reason, now I have more respect for K-8 teachers teaching math. Without some knowledge of child psychology, a typical math PhD may be quickly frustrated in teaching mathematics to K-8 kids.

But summer session is a different ballgame. On the top, the Department is offering courses, for the first time, in a 3-week Summer Session I; 160 minutes per day for five days a week! The time 5:30 to 8:10 PM is geared towards the day-working students. There is a section of MATH 095 in the morning too. Seventeen students registered, but two dropped it before even attending the first class! For students and instructor alike, it is a physical and mental challenge to be in a class for nearly three hours and remain productively engaged. However, students are free to stretch, drink or take a bite whenever they feel like. The session may be akin to immersion technique of learning, particularly languages.

It has been always important for me to understand the students in terms of their background. On talking with another colleague, I asked, "Guess, how many freshmen are there out of 14?" He answered 14; but there are only two freshmen! Besides, there are two sophomores and **eight juniors**. Out of the **two seniors**, one already has a degree, but is taking math courses for teaching certification. The other senior has already walked through the commencement exercises five days ago. She has to take a required core course, **Fundamental of College**

Mathematics (MATH 120) for her degree! She plans to take it in Summer Session II, provided she passes MATH 095.

It is very common for UNLV students to keep postponing a math course required for graduation. It often makes me wonder at the nature of mathematics; fear, anxiety and struggle that a vast majority of students face it. There are some who have taken **Prealgebra/ Arithmetic** (MATH 093) from community college since UNLV does not offer it. In fact, UNLV introduced MATH 095 around 1990. Being remedial math courses, **the credits for these courses do not count towards a college degree.** Students should have learnt this material in high schools, perhaps they did not. Also, math is easily forgotten if not done off and on.

Twenty years ago, the Mathematical Association of America recommended that Precalculus courses belong to schools, and should not to be offered in colleges and universities. The present situation is just the opposite! Up to nine credits in MATH 093, MATH 095 and MATH 096 may be taken before Precalculus/ College Algebra courses.

Fifteen students are spread 14 different majors; **ten have fulltime jobs.** The age varies from 19–45. Three males and twelve females make it a skew distribution. **Three** students are retaking MATH 095! From a feedback survey on the 4th day, I noted that most of them cannot find more than 1-2 hour of study time after the class! **It is a new reality.** Instructors and administration have to be cognizant if such courses are to be offered in this format. It is a challenge in course management and communication. **After all, the proof of a pudding lies in not only in how many students are served in a course, but also in how many finish it.**

In consultation with the students, six class tests of 30 minutes are set for every Wednesday and Friday, and 5 minute quizzes on the material of the previous day. **Yes, a vast majority opted for tests on Fridays rather than on Mondays!** Obviously, thinking is that weekends are not for studying! Overall, course evaluation is flexible and dynamic. **Home work is assigned every day, but not graded.** There is no longer any department assistance in grading papers. At the end of first week, only one student has dropped. The remaining14 have taken both the tests. Student retention is very important to me. **It is robbing the students off their money if more than 50% drop a lower division course.**

These courses provide good bread and butter for faculty, graduate students, and part time instructors. They also generate lot of money for the university coffers. **Everyone is a winner.** Consequently, remedial courses are raging all over the country. **Offer, not to offer, or how to offer remedial courses in all disciplines, is only political for the image of a university.** The demand is heavy and increasing. If UNLV won't do it well, then some other education outfits will do it.

Finally, proliferation of remedial courses in colleges is not dummying of America in mathematics! On the contrary, mathematics is spreading its wings. More and more mathematics is being required by disciplines that did not demand it years ago. For example, thirty years ago, the nursing program required only Intermediate Algebra; today, Precalculus! Also, increasing number of adults are coming back to schools in their 40's and 50's for college degrees, or retooling. They often like to start it from ground zero in mathematics. Though, they are smart in other walks of life.

Please feel free to communicate tidbits and stories on your experience with the remedial courses.

(May 22, 2004)

REMOVABLE IMPEDIMENTS TO MATHEMATICS

On the very first day in all 100 level courses, a short diagnostic test is often given. The idea is to assess students' general working knowledge rather than their ability to recall technical formulas and skills. **My First Day 'Sermon' is that mathematics appears difficult for non-mathematical reasons, like the improper use of worksheets and writing!**

The situation is alarming in remedial math courses. Starting with paper; students invariably divide the worksheets into two vertical halves. With a couple of horizontal lines across further, they divide the sheet into 6-8 equal rectangular spaces for working out the problems. **That is the Don't Number 1**. By limiting a work space, one limits one's thinking. **Mathematics is a new way of thinking**. **The essence of mathematics lies in deductiveness,** where the smallest steps are logically connected. It is muffled by any preset limited space. Never save pennies on paper when doing math.

Back in 1980's, I gave workshops on *How To Do Mathematics* at the annual meetings of Southern Nevada Math Teachers Council. A week ago, it appalled me to notice, that the situation has deteriorated during this period. The way math is taught in high schools, and homeworks, quizzes and tests graded, it all boils down to checking answers.

It baffles to see students hungry for the keys to test problems. In my assessment, correct answers without any work do not earn more than 20 % of the credits. However, one showing details, but fumbling at the end, may earn up to 80 % credits. **It is constantly drilled that in mathematics mere answers are no solutions.**

Dictum: Write spaciously and clearly leaving a blank line in between. Those writing on plane sheets tend to write in noodles, and meander in northeast, southeast directions arbitrarily. The writing jumps to any empty space left out. Such writings are breeding places of mathematical errors. It has little to do with mathematical concepts and techniques. **But students blame mathematics while they are messed up by their own writings.**

Giving a little historical touch to mathematical writing, the education system, introduced in colonial India, is still oriented towards mathematical expression. French is very similar in this respect. Essentially, it is explaining each operation before is it is executed. It is like documentation in programming; non-executable (comments/ remarks) statements are used before a group of executable statements. It tells the author of the code as well as its readers what is to follow next.

Any non-mathematical writing that enhances clarity of mathematical work is mathematical expression. A word of caution; too much and early emphasis on mathematical expression may turn students off already less inclined towards mathematics. But in remedial courses, they eventually benefit when gently stressed throughout the term.

I can't help recalling some established mathematicians, authors and my teachers at Indiana University, Bloomington and Panjab University, Chandigarh who had beautiful mathematical expression. They are: John B Conway, Robert P. Gilbert, Hans Raj Gupta, P. R. Halmos, E. Hopf, and T. P. Srinivasan. Andrew Leonard, an IU mathematical physicist often said: **clear writing generates clear thinking**. Fortunately, a twin habit of penmanship in literature and social studies is still emphasized in schools. It is the US public school systems since 1960's that have gradually eroded mathematical expression.

The third obstacle in mathematics learning is the Yellow Pencil Number 2! Why? Students make a mistake, erase it, smudge that area, overwrite it, and then make newer mistakes! Sometimes the vicious cycle goes on. The use of pencil is a corollary of the scantron grading of multiple choice questions. I discourage the use of pencils at least on tests and quizzes. Those who use ink pens get an extra point!

Its logic is simple; **time is precious on tests, why waste it on erasing**. Also, in life, unless you look into the eye of your error, you are bound to repeat it. On a test, quiz and home work, if a part or entire problem has to be redone, my suggestion is -- gently cross it out, and move to a clean area.

How easily math professionals can make their discipline likeable. All we need to do is to apprise the students of the hidden obstacles and gently guide them in eliminating them. Locally, we try in our own ways, but collectively a global impact can be made.

(May 30, 2004)

ON MEMORIZATION OF FORMULAS

As soon as I announced a test on the material covering techniques of differentiation, a chorus of voices rose up: "Do you want us to memorize all these formulas?", "Can we bring a 3x5 card for the test?", and so on. It happens very often. However, this time, I was in a mood to address it! There being a couple of chemistry majors and quite a few biology majors in the class, I said, "From my experience of **organic chemistry**, one has to memorize far too many carbon compounds. Students taking biology have to memorize hundreds of Latin names of plants, animals, and then strange organic chemical equations."

Continuing, I asked, "Does anyone have a friend studying theater?" **Theater majors have to memorize pages of transcripts without missing a comma, and with emotion**. Once, a theater major told, "It is easy to memorize, otherwise, you are out of the program, if you won't." That is why it is easy! In humanities, one remembers long passages and quotes of authors. I hated history in high school, because I could not remember various dates and events.

A few years ago, I audited a course in music theory. Within a week of the summer session, I felt inundated with so many principles! I knew music theory was mathematical, yet like trig identities, it was overwhelming. Incidentally, a student, who had finished *Finite Mathematics* (MATH 132) from me, was also taking music theory. One day, I asked, "John, tell me which course is more difficult, and requires more memorization?" In a split second, he said, *Finite Mathematics* was easier, and had fewer things to remember!

In the present social culture, there is no shame in being ignorant about mathematics. In a class, students constantly blurt out, "I don't know, I don't know." I constantly drill that mathematics requires minimum memorization, as mathematics is a new way of thinking. It is all deductive. For example, if you remember one trig identity, or one derivative of one trig function, then you can easily derive quite a few.

Instructors need to use strategies to motivate students to memorize basic results. And students do it, if instructors don't relent. If one instructor yields, then students tend to put pressure on other instructors. "God has blessed human beings with this faculty to remember; so use it, or lose it. **After all, your worth and identity are determined by what you remember and what you can recall.**" I do not hesitate to share such homilies with my students.

(June 20, 2004)

A New Look At Home Work

Home work (HW) is the most important component of mathematics instruction. It is very crucial in all lower division (100-200 level) courses as they also lay foundations in many other disciplines too. **The nature and beauty of mathematics are only revealed by doing many problems, and trying one in many ways, if possible.** Experience and excitement of discovery of alternate solutions can not come from attempting sporadic problems, reading worked examples, or watching someone do them.

Currently, the assigned home work problems in lower division courses vary with the availability of a grader and hours for it. Some instructors assign only the odd numbered problems. The reason being, odd numbered have answers at the end! The danger here is that students equate answers with solutions of the problems! I often assign multiples of 3. Its downside is that by and large two thirds of the problems are left out! The students just don't do problems, if they are not assigned. The challenge lies in motivating students to do at least 50 % of the exercises before the next meeting.

During the last 3-week Summer Session I, an idea stuck while teaching MATH 095 (Elementary Algebra) for the first time in 30 years! Every day at least 80 % of the problems were assigned, but not collected the next day. A few typical HW problems were discussed in the class. However, students were encouraged to discuss HW problems with any one. These days, tutorial help in lower division courses is available on some web sites. The students were told to turn in all the home work problems on Friday, the test day. While they took the test, I evaluated and graded their home works.

Integrated with home work was a 5-minute quiz every Monday. The quiz problem was straight out of the HW problems. In lower division courses, the weightage to the home work is 5-10 %. **What surprised me was that the quality of home work improved each week.** Consequently, the class performance improved on tests and quizzes. To check its validity and effectiveness, it was tried during the five week Session II (MATH 181/ Calculus I) and Session III (MATH 283/Calculus III). The results are very positive! Additionally, it saves a lot of Dept money during a time of fiscal conservation.

It reminded me of my college days (1955-59) in India! Neither there was, nor still there is a concept of a grader for a course. The examination system is such that students are responsible for not only all the problems from a particular text, but also from other comparable sources. A few students and instructors are always fishing out for challenging problems. There being a continuous revision of solutions and exercises that by the end of the year, some HW notebooks could be sent straight for a publication. However, college education in India was a privilege, whereas, in the US, it is a right.

(Aug 16, 2004)

A Lot To Learn Some Days

Yesterday, a communication from the Graduate College triggered a recent spate of educational experience. In a span of **four weeks** of July and August, I was present at the MS thesis defense of **four students** in the Department of Mechanical Engineering. Any one from outside the US would wonder at the presence of a mathematics professor on such committees. My membership was not as a subject expert, but as Graduate College representative who is to ensure that the examination is conducted fairly according to the guidelines of the College. Particularly, academic integrity on the part of the student and faculty is held to high standards.

A unique feature of this experience was that all the four students had the same committee members including the thesis supervisor. All four of them worked on the testing of different grades of steel under different conditions of temperature, stress and chemical reaction. I had all theses at least a week in advance. Not, that I could fully understand the work, but wanted to see traces of higher mathematics. There was none! And that is what I was told, when I specifically asked each one to name a particular mathematics course, or topic that they found useful directly or indirectly during their thesis work.

However, I did learn about the time and inventiveness required in the setting up experiments before observations were recorded. Interestingly enough, the data was very small, and no statistical analysis was necessary. For a moment, I was in disbelief that it took 31 days to record an observation. Some students had to visit national laboratories for some testing. **All the students were working on a funded research project of reducing the half life of nuclear material**. It was impressive to learn that their success was mathematically significant. For full safety and commercialization, it still may be decades away. But I got a feeling that with continued research, it would be possible one day to de-contaminate the nuclear waste at the site of its origination. Then Yucca Mountain depository would be closed! I was filled with such thoughts while listening and watching the Power Point presentations.

No less important was to learn what kind of academic work goes on in other departments and colleges. For years, the university departments have been running their own tight ships. But for the last few years, a push for interdisciplinary research in science and engineering is bringing some departments closer. It was fascinating to hear that some of the reported research was one of its kind in the world! This was quite informational. Despite a quality work, the committee members gave suggestions in each case for correcting, adding and editing some portions of theses. At a time when the MS and PhD degrees seem to be on 'sale' in the flight magazines of airlines, it is heartening to see a university system, like the one at UNLV, maintaining rigorous academic standards. See it your self, if you get a chance.

(Sep 03, 2004)

Attendance Optimizes

Last week after the first day's business of giving handouts on course evaluation, syllabus and policies, a dramatic announcement was made at the end of the period. "**Any one with perfect attendance record will be given additional three percentage points.**" It means a student earning a B grade will have B+ as a course grade. The students of **Discrete Mathematics** (MAT 251) were noticeably surprised, as if I was inflating the grades.

Attendance is a big issue in high schools that even tardiness is officially reported. In colleges, there is no general policy and is left to the individual instructors. I have known instructors dropping a student for missing the first day of class, others paying no attention to the absences, and any thing in between. Recalling my hoary days of college in India, it used to be just the opposite. In my **private** high school (90 % then, 80% now), there was no penalty for missing a class or two, but some **corporal punishment waited for the truants**. But in my **government** college (85 % then, 35 % now), the attendance was taken in every period and the absentee fines imposed. At the end of the month, the fine list was publicly posted on a bulletin board!

Last summer, for the first time, I taught a 3 credit course in a **three week Summer Session I**. It was a remedial course in Elementary Algebra (MATH 095). The class met all five days from 5:30-8:15 PM. On the first day all 17 were present, but two missed on the second. The course pace is so fast that unless one has a private tutor, it is not easy to catch up the missed material. It was on the third day, this idea flashed my mind. I told that any one with perfect attendance shall get **5 percentage points**. For example, any one with a D+ at 69% shall get a C at 74 %. **However, one gets nothing for missing even a single class!**

Subsequent attendance was nearly perfect and eventually that translated into a better performance on tests and quizzes. Sample of 17 being small and diversity amongst the students large, I decided to test it on Elementary Calculus I (MATH 181) during the following five week Summer Session II. The class met for four days, Monday through Thursday from 5-7:30 PM. It was a different format, a class size of 34 students and lesser diversity. Though the outcome was satisfying, still I felt it may be singular.

During the last five week Summer Session III was a course on Calculus III (MATH 283) on the same format. It was a group of 47 with homogeneous background. Five percentage points for perfect attendance was announced on the first day. I felt good at the end. Reason: Did it increase daily attendance and course retention rate. **YES** to both. Did it result in grade inflation? NO. Here is the raw data on the Number of students on **Day Number One**/Number that **finished** the course (Class **GPA**). MATH 095: 17/13 (3.09); MATH 181: 34/20 (2.23); MATH 283: 47/42 (2.48). **Side Bar**: At least six students from MATH 095 went on to pass MATH 096 and MATH 124 during summer sessions!

(Sep 05, 2004)

DISCRETE MATHEMATICS AND FOOTBALL

Discrete Mathematics I (MATH 251) and a football game in National Football League (NFL) have a strange vein of similarity. It struck me while grading the first exam given after four weeks. One of the questions was: ***Describe your understanding of proofs in mathematics so far.***

The responses were scattered. Two students said they had no idea, yet they made a fair attempt on three proof problems! It is like in football, that no player can say that I don't know how to play the game. Those who understand the rules and practice a lot, perform better. It is as much true in football as in a math course.

The objective of the course is to impart an understanding of proving and disproving statements in mathematics. It is not a traditional topic course that incrementally builds up from the first chapter on to the second, and so on. Each chapter provides a different scenario for proving elementary theorems and discovering counter examples etc. Usually the topics are real number system, prepositional logic, set theory, number theory, combinatorics, algebra, and graph theory. It may be added, that each topic corresponds to a full fledged math course at upper division/graduate level.

Last Sunday, I heard a sports analyst saying that New Orleans lost to Arizona due to high game time temperature in the Arizona. New Orleans is used to playing home games in their air-conditioned superdome. Suddenly, such remarks converged in my mind. I have seen how the outcome of a game is affected due to frigid temperatures, rains, swirling winds, snow, and drizzles during the game. **The rules of the game do not change with playing conditions.** They remain the same. However, performance of the players is significantly affected when they do not have their home like playing conditions.

Good professional players do not complain and grumble over playing conditions. The coaches train the players in as many diverse and simulated playing conditions as possible. The stakes are very high in each NFL game. The players must adept to the road conditions.

That is exactly is the story of ***Discrete Mathematics***. Each chapter is to be approached afresh. The objective remains the same. Once the basic definitions and symbols of a new topic are spelled out, then it is all about the statements. **Prove it, if true, or find a counter example, if false.**

(Oct 04, 2004)

Mystique Of Mathematics And Madness

Yesterday, I watched a play, *Proof* in Judy Bayley Theater of UNLV. I thought it had some thing to with the famous **Fermat's Last Theorem** that took more than 350 years to prove it. The last mathematician in this multiple relay race is Andrew Wiles. But this play turned out to be different. The focus was neither on a proof of any famous unsolved problem, nor on a mathematician's life. The central theme was genius and his/her being on a fringe of 'madness'. With a cast of only four, the play was very well scripted, directed, and acted.

The general audience can not stomach any one on stage doing mathematics for more than 5-6 seconds. Yet, the dialogues were replete with mathematical terms like iota, primes and complex numbers. At one stage, a Mersenne Prime was rattled off in a breeze. There was a stereotyping that no great theorem can be proved by any one after 25 of age. It is statistically true, that is why the Fields Medal in mathematics, considered equivalent to the Nobel Prize, is only awarded to mathematicians under the age of 40. Nevertheless, in mathematics, like in the world of heavyweight boxing, there is always a George Foreman who wins world acclaim at 40.

The play revolves around a 25 year-old girl who, after the death of her mother, drops out of high school to take care of her father who is a mathematical genius. However, he is often out of touch with every day realities of life. The play has some thematic intersection with the popular movie, *The Beautiful Mind*. Intense and prolonged work in mathematics is like plugging too many electrical appliances in a single outlet. Some where a fuse must go off. In human beings, electrical fields of neuron networks get overlapped. A degree of 'madness' is a function of how complex a fusion has occurred between the neurons.

The theater was nearly sold out. I was impressed with the general caliber of the US public coming out to watch a play that had references to elliptic curves, zeta functions, and theoretical physics in dialogues. Such a wide public interest in science and mathematics is not seen any where out side the USA. No wonders that is reflected in the number of Nobel Prizes won by Americans last week. The American dominance of Nobel Prizes in sciences goes back to 30 years.

The story in the play has a beautiful twist. The father dies, but 'mathematics' genes bloom into one of his two daughters. While taking care of her father, the younger daughter is also drawn towards mathematics, and also pushed by the eccentric father. She herself proves a great theorem. It takes all the drama, romance and intrigue to establish the credit for the proof. What I liked in the play was a non-traditional sub-theme that a woman, like man, can be a genius in mathematics. Of course, she betrays the signs of craziness too! The play has two more showings over the next weekend, and I strongly recommend it.

(Oct 09, 2004)

Is There A George Foreman
In Mathematics?

In an earlier *Mathematical Reflection*, I conjectured that there exists a George Foreman in mathematics who has proved a great theorem after forty. In response, a distinguished colleague wrote back **that it is not possible in mathematics**. George Foreman won the world heavyweight boxing title on Nov 5, 1994, at the age of 46 years and 10 months; a feat that has not been achieved in any boxing weight division. George had gone into retirement in 1977 after losing his world heavyweight title that he first won in 1973 on his 25th birthday. After a gap of 10 years, he started his 2nd successful quest for a world title.

Boxing and mathematics are by and large, individual oriented sports; one highly physical and the other deeply intellectual. But both demand focused mental alertness! Statistically speaking, excellence in each does peak around 24 years of age.

The question that I have is to find counterexample(s) by identifying mathematicians who hitherto unknown, but did a seminal work after fifty. Sure, if there were a Nobel Prize in mathematics, quite a few mathematicians would have won it after fifty as the recent trend in the award process is based on the impact of the work over a couple of decades.

The Fields Medal in mathematics was instituted in 1924, but its eligibility rules disqualify any one over 40. Andrew Wiles, the prover of the famous **Fermat's Last Theorem** in June 1993, being only 2 months over 40 years, was not eligible for the Fields Medal! However, during the 1998 quadrennial International Congress of Mathematicians, a special silver plaque was awarded to him.

It may be of interest to know that the Fields Medal is not annual, but may be awarded to more than one mathematician every four years. Since its inception in 1936, at least two persons have won it every four years. That does make Fields Medal a little more prestigious than the Nobel Prize. However, for lack of a large amount of cash accompanying it, and publicity, it is not very well known outside the mathematics community.

The spirit of America constantly defies challenges in any shape or form. Our city of Las Vegas thrives on visitors who want to beat the mathematical odds in casino games. During my thirty year tenure at UNLV, I had two senior (65+) students who were the best math students in almost all the courses they took. One, Leo Schumann finished his MS at 70+, in 1976. The other Joseph Bechely, nearly disabled by diabetes, enjoyed taking upper division and graduate math courses. Joe literally breathed mathematics and lived for mathematics (or lived on due to mathematics, as he once told me!). Currently, James Brier, an undergraduate at 53 was amongst the top two students in Calculus III that I taught last summer. James aims to go for a PhD in mathematics!

I don't mean to imply that all seniors are smart in mathematics. UNLV encourages seniors to take most courses without any fees. I had quite a few seniors who just could not go beyond Calculus I. However, personal encounters with high achievers senior students lead me to believe that there must be a mathematician (s) whose first fundamental research work was produced after 50. **The focus can be late bloomers and late starters, not on those in the business since their adolescence.**

A correlation between age and creativity in mathematics came into spotlight when the eminent mathematician G. H. Hardy (1877-1947) wrote a classic, *A Mathematician's Apology* at 62. Essentially he theorized that any mathematics research after 25 is of lesser quality, though not linear. It came out of his long and historic collaboration with Littlewood (1885-1977) and Ramanujan (1887-1920), the two great mathematical minds. Hardy, though mathematically active around 40, yet could not keep pace with the bursting genius of Ramanujan, in particular. This classic book came out of his inner frustration to regain mathematical sharpness.

Well, here is a fun deal. Let us identify a mathematician(s) and his/ her fine contribution primarily done after 50. To make it simple, let the time of their life span be confined to the 20th century when all the braches of mathematics had fully emerged. Provide me as much detail and justification as possible by Oct 31. In the selection process, I may take the advice of my colleagues at UNLV. The senders of the top two mathematicians shall receive a cash prize of $100 each.

(Oct 13, 2004)

Keep A Long Weekend Short

Yesterday, it was around 3 PM while walking the hallway, I spoke with a graduate assistant. Being a day before Thanksgiving, I said, "How many students are you expecting in your College Algebra class at 4 PM?" "Not more than two thirds", he said with a bit of calculation. I knew it wouldn't be more than 30%. He also told of having given a take home quiz last Monday. "Then what incentive is left for students to come today?" I added, "I have a bonus quiz scheduled in my Discrete Math class at 4 PM. "

Rarely, I give a make-up test, quiz or exam. That is my policy from the Day Number One. However, there are extra credit opportunities for making it up. The Wednesday's 10-minute quiz was one of the special ones. We walked to our classes. The second floor of the Classroom Building Complex was deserted. The graduate students from a room next to mine were leaving as the professor had suddenly cancelled the class. There was hardly any one in other classrooms across mine. It is nothing new, but only getting endemic.

A class is mutually beneficial when the instructor and students stay engaged through out the course. If an instructor is not on the top of the material, or is unable to communicate it, then very good and poor students will cut classes on such days. Students, not responding to the presentations, would eventually dampen the spirits of an instructor. **However, the onus lies with the instructor**.

A class meeting immediately before or after a long weekend is a different story. Both students and instructors want to have a longer weekend. Canceling a class does have an adverse impact. In a two-day format of class schedule, a typical 3-credit course has thirty 75-minute meetings. **It is not professional on the part of the faculty, if canceled class is not made up**. Yesterday attendance in my class was nearly 100%. On inquiring the students about the attendance in their other courses, I was shocked to hear that mine was the only class that met. The department chairs have to keep a tab.

It reminded me of the 1950's, my college days in India. For the first month of the session, students and instructors had a 'tune in' period for lack of text books, schedule, and host of other reasons. Looking back, it was all wasted time and energy. The system was such that the students who wanted to learn would come to classes, and the professors, who loved teaching, would teach. It was acceptable for instructors to walk in a class 5-7 minute late and dismiss it 5-7 minute before time. Later on, I learnt that former British colonies had laissez-faire educational policies.

In 1968, while at Indiana University, I had a modern abstract algebra course scheduled for three days: Tue, Thu and Sat. No one missed a Sat lecture of Elizabeth Hornix, a visiting assistant professor from Holland. The academic ethics are declining since 1960's.

As a professor and associate dean, I feel an obligation to at least slow down this slide. A new academic policy may not be called for, but awareness amongst the faculty can bring a transformation.

Students at UNLV are absolutely non-traditional. Yesterday, Jack, a 75-year old perennial and former student of mine, walked into the middle of my lecture and occupied an empty chair to the chagrin of my students. Jack told me that his math class was canceled, and on finding me teaching, he entered the classroom!

(Nov 25, 2004/Thanksgiving Day)

AFFINITY FOR MATHEMATICS

By any count, 120 students doing master's in mathematics from a college with 3500 students, is an eye opener on the popularity of mathematics. The Holkar Science College, Indore, in Madhya Pradesh, a central state, was started in 1891, as the first science college of India by the erstwhile ruler prince of the region. The colonialists only encouraged the study of British literature, history, philosophy and economics. Mathematics, being free of national origin or geographical boundaries, was an exception!

I happened to be in Indore as the College had hosted a four day (Dec 16-19, 2004) International Conference on the History and Heritage of Mathematical Sciences. The Holkar Science College, like several other colleges in the city, is affiliated with Devi Ahalyabai University of Indore. What impressed me most was a large number of graduate students volunteering in the organization of the conference despite their final exam week. They even put up an exhibition together on the historical landmarks in mathematics.

Here is the gist of the data. Nearly 60 students finish their master's each year! Mind it, in Indore, a city of about 3 million people, there is at least one more institution that offers master's in mathematics. Another surprising data was that **80% are girls!** One naturally wonders what do they do with master's in math? Only 5-6 students can go for M.Phil or PhD in mathematics. There are not enough research scholarships and fellowships for doing math PhD's. When I got my masters in 1961, Panjab University Chandigarh had only one PhD scholarship! Eventually, I came to Indiana University in 1968 for PhD.

A few days later, I gave a general talk to math graduate students and faculty of MSU (Maharajah Syaji University), Vadodara (Baroda), another 3-million strong city in a western state, Gujarat. MSU is a typical comprehensive state university started as a college over 100 years ago by another princely ruler of the region. The yearly intake is 80 students in a typical two year master's program. However, the graduation and gender data were almost identical, though the two institutions are 350 miles apart!

The math curriculum in both institutions is very structured in the first year, but flexible in the second. Also, it is fairly balanced between the traditional areas of pure and applied mathematics. Most of the students go after lucrative positions in banking, insurance, management, and administrative services of state and federal governments. The private business of tutoring in mathematics is getting very popular. **The general belief is that mathematics is far more helpful, no matter what else is further studied or pursued.**

It is amazing to observe that during all my 30 years at UNLV, hardly 60 students have earned their master's in mathematics! In contrast, one institution in India is turning out over 60 graduates in one year. But any comparison stops right here.

(Jan 09, 2005)

WHY CALL IT CALCULUS?

Yesterday, being the first day of class of *Elementary Calculus I* (MATH **181)**, I thought of starting the course with a little salsa and bang. A question was posed, as to why this body of mathematics is called Calculus. **Very often the dictionary meanings of mathematical terms are reasonably close to their mathematical definitions that only add preciseness and eliminate ambiguities.** However, the word calculus has posed a sort of challenge. The *Webster Dictionary* has one meaning: *An abnormal stony mass or deposit in the body.* The *American Heritage* says: *An abnormal concretion in the body usually formed of mineral salts, a stone, as in the gall bladder, kidney, or urinary bladder.* The other is a **method of calculations,** and the rest are on its usages. Calculus is a singular noun; its plural being calculi, or calculuses!

My immediate reaction to my own thought was that how come I was still not sure about it? After all, I have taught Calculus since 1961! Did it never occur before, or have I forgotten a story behind it? I confessed it in front of my students adding that I never kept notes of my lectures. However, now I am going to keep some!

I tried hard to connect the dictionary meanings of calculus with topics that are done in calculus courses. **The spirit of Calculus lies in the concept of Limit.** Incidentally, dictionary meanings of limit are conveyed by the $e - d$ definition of Limit! The Limit has infinity and ad infinitum built into it. Whether a bunch of numbers approach a number or values of a function approach a number, the process has an inherent movement of infinitesimal steps. **Isn't it akin to the formation of a glandular stone?** A stone formulation is a very slow process of accumulation of minerals around a super micro singular point in a gland. Only after a long period, it becomes noticeable. This is called a limiting process in Calculus! Well, that is an explanation on Calculus I gave it to my students.

The word calculus being of Latin origin also means *small stone used in reckoning*. That may be in dark ages, but it is calculating nevertheless. On talking with a colleague last night on this subject, he made a point about the nature of early mathematics. Calculus emerged out of two historical problems. One revolves around the slope of tangent at a point on a curve and the other on the area bounded by a curve. Mathematics till the end of 18th century was essentially calculating more of quantities and less of proving theorems. Hence the word **calculus is connectable with calculation.**

However, I declared this issue as a topic for an extra credit 300 word writing assignment. Students will do library research, or go on internet to **find out when the term Calculus was first used in mathematics and why. Did Newton or Leibnitz call a bunch of mathematical topics as Calculus?** I look forward to glean some interesting nuggets about Calculus from these papers, though none of them is going to be from a math major!

(Jan 20, 2005)

IT IS MORE THAN THE *NUMBERS*

Yesterday's second NFL game not living up to my expectations, I decided to 'balance' it by watching the premier of the new CBS show *NUMBERS*. The frequent references on the applications of higher mathematics in solving a case of serial murders intrigued me enough to stay on in front of the TV. It was an hour long episode.

Without question, the show takes the popularity of mathematics at a higher level. It is not surprising either. No where else in the world is seen an awareness and appreciation of mathematics amongst the general public as in the US. It is steadily growing. For example, thirty years ago, one could graduate from UNLV without any course of mathematics. Not any more. Without exception, everyone has to take at least a comprehensive course on the fundamentals of mathematics. The disciplines that required rudimentary courses in algebra or Precalculus have gradually increased their ante by six credits requiring solid courses in statistics and calculus.

Many a scenes in the Show reminded me of the movie, *The Beautiful Mind* (2001) on the life of a mathematician, John Nash who won the 1994 Nobel Prize in Economics where his mathematical results first found breakthrough applications. In the Show, mathematical formulas, modeling the crime data and its dynamic modification with new data, or an anomaly, conveyed the power of mathematics.

The bottom line is mathematics can solve any problem, provided one is persistent and knowledgeable enough. The Newton's Law of Cooling has been used to ascertain the time of murder for over a century. One can recall how the LA city coroner was grilled on this Law for two days in the O J Simpson Murder Trial (1994).

The Show began with general remarks on the well known applications of mathematics. It quickly mentioned predictive theory, inverse problems, and theory of patterns in randomness. The producers and directors went on a limb in weaving the story of a great 19th century mathematician Ernest Galois into the episode. In a story line, they also reinforced a 'fact' that the creativity in mathematics peaks before 30.

Particularly relishing were references to mathematics of String Theory and Gravitational Forces stemming from Einstein's last work on the Unified Field Theory of the universe. It is no coincidence, as the year 2005 marks the centennial of 1905 (**annus mirabilis**) when Einstein published 4-5 papers that has changed the course of mankind for ever.

The academe can capitalize on such events in the media. It is only through their integration in math courses that dividends can be reaped. It is time for **math instructors to play some role of entertainers while teaching courses particularly at 100 level**. Let us keep this show on the air by watching it, and encouraging our friends and students too.

(Jan 24, 2005)

LET X = X

Is dance mathematical, or mathematics a dance form? In the mind of at least one person, Cathy Allen, Associate Professor of Dance at UNLV, it is, perhaps, yes for both! As soon as a university wide announcement came out on her choreographed dances including one mathematically named, *Let X = X*, I gave her a call. Yesterday, my wife and I watched a 50-minute performance of six dances. They were fast, energized and well synchronized.

The *Let X= X* was premiered in 1992 in LV Choreographer's Showcase. In a brief introduction before the performance, Cathy, in the background, said, "It is dance about my imagination during 10th grade math classes. Needless to say, I didn't receive my best grade in that class. Later on, when I was matured, then I did well in *College Algebra*."

One may right away ask, 'where is the beef?' type of question on the presence of math as done in typical math courses. Math submerged in ever changing patterns of dance movements requires a special eye to see it. **Cathy's memories of her 10th grade math were so strong that after 20 years, it erupted as her creative work!**

For Cathy, or for any accomplished dancer, dance becomes a window of life, a paradigm. It is like mathematics for me. The speed with which the dancers, oblivious of the audience and other dancers, run, and move backward, forward without any collisions, is because of mathematical precision in their foot work. One has to focus on the movements of hands and feet to observe how they all synchronize. **It is a visual geometry.**

Yes, integrated in one dance were high-tech projections and mathematical symbols as one sees in a new CBS show, *Numbers*. On the stage were ten chairs for ten 'students' who danced their math out! It was an audio-video doodling. Every one of us has done it with pen and paper during moments of boredom. Call them lapses, or oozing of creativity!

If a professional mathematician would look at a proof of a theorem from the perspective of a dancer, then it surely would fall into a dance of mathematical symbols! **Not every body movement is a dance movement. Like wise, any set of mathematical symbols does not make a mathematical statement.**

In a dance, a choreographer has to work with a light engineer for various hues to fall at the right spots and from the right angles, with a costume designer for color and expression, with a sound engineer for audio effects, and with music director for melodies for various moods. **They all together capture the soul of a dance.** A mathematical work, by and large, does not depend on any thing other than austere paper and pencil. That is a stark contrast. **Above all, it speaks high of the US society for its growing appreciation and acceptance of mathematics in newer walks of life.**

(Jan 29, 2005)

NEW CHALLENGES OF TEACHING

At UNLV, the undergraduate courses are offered at 100, 200, 300, and 400 levels. The challenges of teaching the lower division 100-level courses are no longer simple. It does not matter whether you have years of experience and polished notes. The growing diversity of students throws off any model of instruction. **It is noticeable only when an instructor is sensitive to the content of his/her lectures and presentation.**

About instances of sensitivity; an operator of a restaurant goes to any length to make sure that good food is prepared, and also served right. Every entertainer makes sure that his/her script is honed, and perfectly delivered. Lately, that has been motivating me in instruction. After all, 80 % of the revenue is generated by 100-level courses. I call them bread and butter courses at times, or trench courses, at others.

Yesterday, I quizzed the students of **Elementary Calculus I** (MATH 181) on the following problem: Find the

$$\lim_{x \to 4} \left(\frac{\frac{1}{4} - \frac{1}{x}}{4 - x} \right).$$

It was similar to an odd numbered (**answers given at the end**) problem collected as one of the home work problems only two days earlier: Find the

$$\lim_{x \to -4} \left(\frac{\frac{1}{4} + \frac{1}{x}}{4 + x} \right).$$

28 students out of 37 (Total 40) did not get it right. Primarily, they made all sorts of errors in simplifying the rational expression. **Yes, only 9 got it right.** It may be added that **3 students abstained from a 100-minute biweekly class**.

My first reaction was commonly, that the students were not prepared and had no prerequisites. On checking it out, the story was different. At least 27 had a prerequisite course, and **10 out of 28 had As in those courses**! The rest had Bs, Cs, or a qualifying score in SAT. That poses a dilemma. How to explain, and rectify this situation?

The present day students are like weight watchers. If they decide on exercising, then they would lift weights for an hour, run 5-7 miles on the first day, and then feel sore for a week before quitting. If it comes to cutting on calories, then the approach taken is to starve to a point that the eating is resumed with a vengeance. **Mathematics is like eating apple a day. If students take even a day off from class material, then it starts looking stranger**. After a long gap, one may be a bit wiser in disciplines, not heavily dependent of mathematics.

Such a performance happens if at least 60 % of exercises are not done in prerequisite courses. The high school practice of attempting only 10-20 % of textbook problems continues in the college. Mathematics does not stick long enough. It evaporates especially when the students treat the courses as obstacles to be taken 'out of the way'.

There is another resemblance between learning math and getting in shape physically. One does a few stretching and warm up exercises before a game or a run. During advising and at the end of a math course, it needs to be emphasized to the students that they must take math courses without any gap, and give a head start on the material at least a week before the classes begin.

Students at UNLV are unique in many ways. Quite a few have family responsibilities and have full time jobs. In my section, 20% of the students work full time and take at least 12 credits; 40 % have part time jobs and carry at least 12 credits, which is a fulltime credit load at UNLV. Though, it is a freshmen course, only 25 % are freshmen, and **at least 25% are repeating this course**. On a bit personal note, there is no math major in my section (total 10 sections)! The average age is nearly 30, very non-traditional group.

Talking of family responsibilities, it has been an American practice that while one spouse goes to college, the other works and puts him/her through. No one demanded a special treatment for being a parent. With high schools accommodating unwed teen mothers, providing mothering facilities in school premises, and opening alternate schools, now students with families and jobs essentially force 'lighter' college courses.

Regarding real special treatment, there are a growing number of students with genuine special needs. The number has quadrupled in ten years. Instructors have to accommodate them with alternate tests, more test time and coordinating with different test centers.

The diversity in a litigious environment raises pertinent questions on course management. Based on the data of previous offerings, I do tell students the odds of finishing a course with various combination of working hours and credit hours. Even an academic advisor cannot compel them. The ultimate decision is of the students. **To a large extent, such overstretched students represent the spirit of America in defying the odds!**

What is the consequence of this approach? Last summer, in my section of MATH 181, nearly 50 % got a C or higher grades. The rest of the students either dropped the course at some point in time, or had a D or F. The students no longer have any qualms in repeating a course semester after semester.

It is unlike my times of 1950's in India, when if you flunk an exam, then you go through the motions for one year before taking it again. I know people making four attempts, meaning taking four years to pass a certain exam! Is it some tyranny of an education system? **Did those flunkeys end up at the bottom of a social ladder? Not at all!** A few made big time success in business and politics! Any college, whether in the US or India, can only prepare for 8-10 specific life exams. **Life poses infinitely many challenges.**

(Feb 03, 2005)

New Course-Burgers

Things that happen in the world of mathematicians are not commonly heard. Two days ago, eighteen new graduate courses were proposed. Though the MS Program in Math is 35 years old, but on the average only three students finish it per year. Currently, two out of three go in applied statistics. The PhD program, approved in 2004, is on hold till 2006.

My thoughts went to the fast food giant McDonald's! They release a new food item after every 4-5 years. The single most important thing is a continuous survey done on the trends of food tastes. Once that indicates a futuristic item, they go into its every aspect of research from developing the right meat to other ingredients. It takes a couple of years of lab work alone. I know it from a personal account of a research scientist with KFC.

The futuristic item remains under a shroud. After the lab results are satisfactory, the actual taste survey on potential customers follows. We all have tasted bites of new food items set up on small tables in the middle of prominent areas of grocery stores. McDonald's particularly may not go through this route, but they do run a taste survey to find if further improvements are necessary. Sometimes the item is completely withdrawn and started from scratch. The bottom line is testing and tasting going at different levels before an item is finally released on prime time TV with fanfare.

The story of mathematics courses is no different. **As far as the new courses are concerned, a demand has to be created and established**. During the last 30 years, the number of undergraduates graduating with math major has not touched even five in one year! This number is likely to go down due to additional math requirements effective fall 2003.

Once a demand for a particular course is established, then first it should be offered as an experimental course or under a generic course tile of **Topics**. That is the best approach to debug the syllabus, develop notes, and refine the material to meet the background of UNLV students. Above all, it establishes integrity of the course and the instructor.

What is the point in investing time and energy in a course(s) that will never be populated with students? There are already at least ten graduate courses in the catalog that may not have been offered even once in 30 years! Not all students majoring in math go for MS, or PhD. **Unless faculty turns its focus in correcting the undergraduate situation, the creation of new graduate courses will be like proving theorems in empty sets.**

(Feb 12, 2005)

WHERE IS A LIMIT?

Sometimes a truth is stranger than a fiction, no matter what walk of life it may be. It was Friday afternoon, as I returned from a meeting. "Satish, call me as soon as possible." was a message on the answering service. It was from an Associate Dean of the College of Engineering and Director of Minority Engineering Program. We often discuss academic difficulties of the minority students at least once a semester.

"You won't believe this story." That is how he spoke up, when I called him right away. I said, "I have heard it all in over 30 years at UNLV." "But this one tops it all, I guarantee it." He was right, now you decide it.

It is about a civil engineering major who took STAT 463, *Applied Statistics for Engineers* as a graduation requirement. The prerequisite for this course is MATH 283, third semester calculus. **However, this student was enrolled in MATH 096 (Intermediate Algebra) at the same time**! Forget calculus, he did not have any Precalculus or Intermediate Algebra! Yet, he finished STAT 463 with a C Grade!

In MATH 096, he had a B; dropped MATH 126 (Precalculus I) twice before finishing it with a B in the third attempt. Currently, on facing problems in MATH 127 (Precalculus II), he approached for getting a tutor. That is how this story divulged.

Since then host of thoughts have been ricocheting my mind. Most students do flock to instructors who give easy exams and high grades. Four years ago, a graduate student complained against me for his B grade. Perhaps his all other grades were As. Last fall, the grade inflation in the College of Sciences was identified in 100-level courses.

I hope this story is a limit. The student must have failed every HW, quiz and test, and yet got an average passing grade! He must have spread this word. One can discuss its impact on the image of an instructor, program and UNLV. I have encountered such students eventually getting not only BS, but also an MS!

They perfect the game of choosing the right courses and instructors. There are more than a couple of students lined up for upcoming Math PhD program. How can you deny admission when they flaunt your As and Bs on their transcripts and MS diplomas?

These days, no one wants to face a grievance and a law suit. Alternative is to steer such students out through the maze of a program they are admitted to. The problem is serious; it is national and international with all the online PhD programs even in Math.

Some believe that real academic integrity begins and ends with an individual instructor. Some believe a peer pressure alone can correct it. I may be cynical, as I have not seen it abating in my 30 years at UNLV. Reason: Teaching remains at a short end.

(Feb 13, 2005)

It Is More Than Cost Effectiveness

If an incident does not erase from the memory after a week that means it may be a nugget of an experience. It happened a week ago when I visited a middle school as a **PAYBAC** (acronym for **P**rofessionals **A**nd **Y**outh **B**ring **A** **C**ommitment) speaker. The **PAYBAC** is a community and school district partnership program to encourage kids to stay in schools. Only minutes before the lecture, I was told that the period of the 6th graders was of mathematics.

The topic of the speech does not have to match with the subject of the class period. But I enjoy the challenge of making a connection with mathematics at a short notice. The 40+ years of professional life so far has not been a repetition of only a couple of years.

After the introduction, I asked the kids, **"What would you do to make your body strong?"** Involving all 31 of them, they came up with 11 activities including weight training, most major sports, gymnastics, swimming, and running. All were listed on the board and I put an asterisk in front of only 'running'. No one could figure it out as to why I did it. I said, "Running is the only activity that costs hardly any thing!"

Adding push-ups and sit-ups to running, I told the story of Herschel Walker, the legendary NFL running back with perfectly chiseled physique. As a junior high school student himself, and on following the advice of a high school football coach, he developed this physique by doing only push-ups, sit-ups, and running fast around his block during the TV commercial breaks.

Having captured their attention, I also told them if they wanted to develop the 'logical' side of their brain, then they should study mathematics! Look how much it costs in studying sciences; dissection experiments on live creatures in biology, materials in physics, and chemicals in Chemistry etc. Math is still done with paper and pencil. Also, mathematics is a language of sciences. Personally, I studied math as there were no science courses in my college during the last two years of my bachelor's back in India.

Doing arithmetic calculations in the head invigorates brain cells. That is called mental calisthenics. There are many research reports supporting it. In conclusion, the kids did a few 2-digit multiplication problems like 51 x 49, 53 x 47 and 55 x 45. It was not easy for most of them. Some resorted to the use of calculators. Then, I explained it as an application of factorization of the difference of two squares. The kids were amazed as they all could do such multiplications in their heads.

At the end, I gave out a secret that all these multiplications come out of an application of algebra that they would learn in the 7th grade. The teacher listening and sitting at the back of the classroom smiled at this pitch for algebra.

(Feb 17, 2005)

HOMEWORK, A DAILY RITUAL

"Hey, there is no home work assigned today!" was one of the varied expressions of glee that I overheard while cleaning up the board at the end of the period. On some days, specific homework problems are not assigned to be turned in, as was the case yesterday in **Calculus I** (MATH 181). Also, a few e-mails of inquiry came from the students: "I didn't hear you assigning any homework due for Wednesday. Please let me know if I am wrong."

Home work is a popular American phrase meaning going prepared for any meeting, or presentation. The weightage to the homework usually starts from 5 % in remedial and 100 level courses, and increments to 20 % at upper division 400-level. In all the courses, my approach is to break the high school mentality; if no home work assigned, then there is no studying at all before the next meeting.

In response to the e-mail, I wrote, "Hey, all the textbook problems are yours! Get your money's worth by doing each and everyone." It is not 'vacationing'. Mathematics is a unique discipline. Whether one is student or instructor, if the problems are not tackled every day, then gradually that area of mathematics starts slipping away. Personally, a few courses, taken as a graduate student more than 30 years ago, are literally out of my system, as I have not taught that material, or used it in specific research.

Whether, homework is assigned or not, doing a few exercises daily in a math course is like an apple a day for good health. On Day Number One, in the syllabus, important problems from each section are identified. They must be done for any claim on the understanding of calculus. **Also, it is stressed that for every hour in a math class, one must study for at least two hours on the same day**.

Under the system I studied in India during 1950's, problems were not assigned, collected and graded! A few students did every problem from the textbook, and then hunted for the challenging ones from other sources. We were more than full-time students, as there was no other responsibility as today's students have it of jobs, independent living, or family. The average age of UNLV undergrads is 28. **Though Calculus I is the first college course, but only 25 % are freshmen in this course and nearly 100 % are working; part or full. No wonder, 33% are repeating this course!** Repeating is no longer a stigma. Gone are the days when one would finish a course in the first attempt.

The Indian system of final examinations also forced a different attitude for studies. On coming to Indiana University for PhD, I followed the same approach; doing all the problems whether assigned or not. It paid off. In mathematics, one needs to practice on drill problems as well as on a variety of them. **To a large extent, this applies to lower division courses in any discipline.** They are fundamental and foundational, irrespective of the major.

(March 08, 2005)

Philosophy Turning Mathematical

Yesterday, I attended a colloquium given by David Sherry, a philosophy professor from Northern Arizona University. The Title of the talk, *Bayes' Theorem and Reliability*, and a line in the Abstract, '*Numerous psychological experiments appear to show that people are not very good at inductive reasoning,*' caught my attention. I think the concepts of induction in mathematics and physics are very intuitive.

Currently, this issue seems to be hot in the law courts and clinical psychology. For math students, the scenarios of word problems are not important while applying mathematical principles to the problem. We let the chips fall and move on to the next word problem. The context of the problem is absolutely irrelevant. No human values or judgments are called upon in solving a mathematical problem, or in analyzing its solution.

Bayes Probability goes back to Thomas Bayes (1702-1761) a British **priest** whose father was also a **priest**. Sherry's lecture was on conditional probability and Bayesian Formula. In a finite math course this material is covered in two lectures. Two scenarios, that Sherry presented included one from NBA and other court witness, raised questions of reliability. In the domains of philosophy and psychology, it often becomes a ping pong of ideas. **It happens because the heart does not accept mathematical solutions, and the head is not equipped to discover new mathematical tools.**

During the lecture, my thoughts were bouncing back and forth. The first observation was that **the western philosophy is no longer what its Greek heritage conveys**! Twenty five years ago, I attended a philosophy colloquium delivered by a vice president of an international organization of philosophy on a topic of 'identification'. He made its far fetched connection with an integral equation! Since then, the study of philosophy has become highly mathematical.

The colloquium attracted nearly 30 people on Friday afternoon. One half of the audience was from psychology and the other half from philosophy. The funny thing is philosophy and psychology departments at UNLV do not require a serious math course beyond a baby course on the fundamentals of mathematics (**MATH 120**) taken to satisfy the University's General Core Requirements. In order to keep up with a growing mathematical usage, at least 1-2 courses including calculus should be recommended. The understanding of deductive reasoning of mathematics is very pertinent.

Thirty five years ago, a well known Indiana University (IU) psychology professor gave a colloquium in the IU Math Dept. He applied the concepts of set theory and mappings in modeling an area of psychology. I was then a recent graduate student from India; it was a breathtaking experience for me to see such an application of math. In the audience were mathematics legends like Zorn, Hopf and Halmos who had left their mark in this field. Since then, Mathematical Psychology has matured into a respectable discipline.

The inroads of science and mathematics, started in the 15th and 16th century, propelled western philosophy into their mathematical orbit. The western philosophy, in particular, the way it is approached in the US can be described as highly utilitarian. At the other extreme, the Hindu philosophy is totally divorced from life.

Being a member of UNLV Asian Studies Advisory Committee, I happened to meet a University of Michigan Professor of Buddhist Studies over a group luncheon last year. There, a colleague remarked, "Is there an Asian Philosophy?" In fact, it was meant to include any Asian country! Surprised to hear this, **I simply said that the Hindus have gone to the extreme of philosophizing every thing in life!**

There is a gulf between what is considered philosophy as a discipline in the US and what it is in India. The questions relating soul, heaven and earth, mythology and symbology, ethics and morality integral to Hindu Philosophy are inseparable from religion. They have been out from the domain of the western philosophy since the 18th century.

Another thought was that if I have to pick up **one feature that distinguishes the American Civilization from the rest of civilizations in history, then in one word, mathematics, says it all.** Today, even serious mathematics is percolating disciplines hitherto were light years away from math. I feel good about my own observation!

Don't get me wrong, I am not for more math. However, I am for the right use of math. Math being binary in nature has more limitations than literary languages like English and Sanskrit have them. **Math is rational in approach while philosophy is supposed to be non-rational.** Mathematization of a problem only provides a very different perspective that necessarily may not be the best.

There is a downside of math visibility. For example, a few years ago, a colleague in the Criminal Justice Dept. showed me two manuscripts of the same paper. One, having no math, was rejected by a journal. However, the second one, re-submitted after including a table of data, was accepted by the same journal! Since then, the 'abuse' of math has gone very far such that new courses and disciplines are mushrooming up in academia.

In the US, the pursuit of philosophy is at a bifurcation point like economics divided into quantitative and qualitative. There is a 'theorem': **Social Studies + Mathematics = Social Sciences.** Amongst mathematicians, there is a growing interest in the philosophy of mathematics. A few years ago, a special interest group under the umbrella of the Mathematical Association of America has been actively promoting it. The website maa.org has details. **Philosophy of mathematics that philosophers removed from mathematics are engaged in is not the same as discussed by mathematicians and philosophers that are at least knee deep in mathematics.**

Summing up this entire philosophic exercise, there is nothing like the life of an intellectual in a US university! It provides an ultimate buffet of ideas.

(April 09, 2005)

On Short Papers In Calculus

Literary writings and doing mathematics are generally considered poles apart. As mathematical sophistication rises, the number and complexity of notations increase. For getting an idea as to how advance math books look like, one has to scan Russell and Whitehead's Magnus Opus on logic, *Principia of Mathematica*. Hardly an English sentence appears in pages after pages.

The present state of mathematics in the US is very interesting. One may get pessimistic on not finding enough math majors. For example, in my sections of Calculus I, there has not been even a single math major during the last two offerings! The positive side is the growing number of mathematical applications, as mathematics continues to conquer new territories. Consequently, math and statistics requirements are increasing in all disciplines in business, liberal arts, fine arts, and social studies. It is good for math business!

Two years ago, while teaching Calculus (HON 141H) to non-science majors in the Honors College, I introduced written reports as homework for the first time. Writing is integral in the Honors curriculum. I discovered writing about mathematics was motivational.

Since then, short writing assignments have been tried with other classes with positive outcomes. It is just like after going to a movie, restaurant, or musical, we eagerly like to share its good and bad experiences with our friends. So why not try the same with calculus in writing? **Any thing, when put in black and white, registers deeper in the mind.**

Calculus I (MATH 181) is the story of three fundamental concepts of mathematics, namely, Limit, Differentiation and Integration. The students are suggested to write each report in the form of letters to their dear ones, who may or may not have any idea of Calculus. In letter writing, the focus on one person brings the ideas together more truthfully. It is fun to read them. Varied influence, during the formative years of learning, goes a long way.

There is a fourth paper relating the entire contents of the course with any other subject– be it a hobby or a major. **It is a capstone assignment, an assessment tool of the course that is due at the very end of the semester.** So far, these assignments have been optional and counted for extra credits. In the computation of final grades, it carries a marginal weight of 3%. More than 90% of the students turn in at least two of them. Next time, these four reports would be a mandatory part of the grade.

(April 16, 2005)

LAS VEGAS & SAUNDERS MACLANE

Today's early morning e-mail from a neighbor prompted me to click on an internet link to an article in the **Boston Globe**. Its heading read: "**Saunders MacLane, developed key Algebraic theory; at 95.**" I said to myself, "My god, this man is doing fundamental research in mathematics at 95!" But this reality lasted only for a second. In his death on April 15, 2005, MacLane has created a record as the longest living and active mathematician of the 20th century!

While replying to my Bostonian neighbor, I said, "Maclane is the man responsible for not letting the Joint Annual Meetings of the AMS and MAA take place in Las Vegas! Very stubborn man to the end."

I continued to muse over my comments. How can you crack a long standing mathematics problem or develop a new concept without persistence and perseverance? Never! One's being stubborn in dealing with colleagues may not be a positive quality. Nevertheless, in certain areas of 'pure' mathematics, there may be a correlation between originality and stubbornness. It is not a litmus test, though in my small sample, it has worked with 95% accuracy!

During the first and only Joint Meetings held in Las Vegas in Jan 1972, MacLane pushed a resolution that has ruled Las Vegas out as a conference site! I was shocked to discover it, when my interest in the Meetings grew in 1980's. A couple of math bigwigs essentially told me, "You can forget Las Vegas as long as MacLane is alive!" Once I wrote him a very persuasive letter on this issue. He graciously replied it in his long hand, but refused to change his mind.

One may talk of his legacy; both mathematical and non-mathematical. **They are two separate things.** One does not have to study **Category Theory** to know its impact on many branches of mathematics including computer science. His influence on mathematics organizations is deep for the number of people who know him directly (his colleagues and PhD students at Level 1) and indirectly (his students' students, Level 2, 3, 4, and 5!). To give an idea of his hold over math organizations, I never got a satisfactory reply from any officer of the AMS/MAA of my written suggestions for a Joint Meeting in Las Vegas! It was essentially a one man crusade that I loosened up five years ago.

Yes, it was five years ago, when I last saw MacLane at a Joint Meeting. He spoke at a panel discussion on *Philosophy of Mathematics*. People were standing wall to wall in a big hall to hear him! He had notes written and rolled up in yellow sheets. Most of the time, he rambled on. **In fairness, that was a great mathematical performance at 90!**

MacLane must have been asked a secret of his longevity. I think, had he not re-married at near 80, he would have died of loneliness (A line from a work of George Bernard Shaw)!

(April 23, 2005)

Course Organization 101

Finally, last spring, I lectured the students on how to keep the course material organized. It is a 'pre-requisite' for learning mathematics particularly at the lower division level. About the same time, a discussion had started on a new 1-credit course on '**Orientation to College Studies**' for incoming freshmen students in the College of Sciences.

For making an impression, I brought out a three ring folder divided into **seven** sections. Seven is universally a favorite number! The **first** section is for filing important handouts given on the first day of classes. The students must be aware of the evaluation criteria all the time, policies on attendance, make-up of homework, quizzes and tests. They also need to know the policies on plagiarism and incomplete grades, in particular.

Section **Two** is for class notes. Taking notes is not human tape recording. Jot down the salient points that may help in understanding or to remind you later what was missed during the lecture, or to be emphasized later on. **The important thing is to review the class material first before attempting the homework problems**.

Section **Three** contains all the assigned homeworks graded or upgraded in **a sequential order**, identifying challenging problems. They alone need to be reviewed before a test or quiz, instead of going over the entire material in a shorter time. Homework is the most important component of mathematics learning. This section may contain additional problems attempted. If an exercise set has 70 'eligible' problems, then all the 70 are not assigned, collected at the next meeting and graded. If, say, 30 are assigned, then the remaining 40 are not to be forgotten! **They must be attempted during the course of the semester and their worksheets properly filed**.

Section **Four** is for the graded quizzes and their solutions. A quiz problem(s) is/are usually closely related with the homework problems and the ones discussed in class.

Section **Five** is marked for the graded class/short tests and any information on review material and their solutions. Some instructors get the tests back after letting students see them. So far, I have not done this, though the Final Exams are retained for six months.

Section **Six** saves extra credit, assignments on homework and written reports.

Section **Seven** has interesting mathematical pieces like my specific *Mathematical Reflections*. This material enhances mathematical awareness, interest and thinking beyond the confines of the classroom, course and degree program.

While writing this Reflection, I was reminded me of a popular book, *Whatever I wanted to learn I learnt in Kindergarten*. Yes, I did learn a good part of **Organization 101** at the junior high school level back in my hoary days of studentship in India.

(May 25, 2005)

A Changing Face Of Mathematics

About twenty five years ago, the Mathematical Association of America recommended that the courses below calculus should not be offered in the colleges and universities. Reason, they belong to high schools. Later on, it also got the support of the American Mathematical Society. Being fresh out of the graduate school, I too believed that PhD faculty, regularly teaching courses like intermediate algebra and precalculus, was an under-utilization of their PhD degrees. The heart of PhD faculty generally lies in teaching upper division/graduate courses, or/and continuing with doctoral research.

Today, my one thought on the state of mathematics is, how wrong was the vision of the professional organizations, and why! Mathematics organizations are generally dominated by research oriented faculty. And research, be that of any quality or quantity, matters in all personnel decisions in most colleges and universities. On the contrary, no one then imagined that precalculus courses would be thriving after 25 years! Of course, the term math remediation did not exist in the curricula then.

Last week, UNLV Provost called a short-notice meeting on the problem of remedial courses in English and Mathematics. This problem is being addressed at a Nevada System level. I was there at the meeting attended by two college deans, associate provost, associate deans, chairs and directors. That gives an idea of the magnitude and importance remediation is drawing from the university regents and system chancellor to all the way down below to departments!

The College of Sciences Dean immediately appointed a faculty member as the coordinator of math diagnostics tests. A large number of lower division courses are seamlessly transferable between four community colleges, one four-year college, and two universities in the State. It is essential that the diagnostic tests, given to place students in remedial and precalculus courses, are comparable. That will involve periodic consultation between institutions, administration, and review of diagnostics tests.

This scenario is not intended for any institutional generalization. The focus remains on institutions like UNLV situated in a growing cosmopolitan and diverse city like Las Vegas. After living here since 1974 and traveling around the world, I can confidently say that there is no city like Las Vegas on the face of the earth today. It turned only 100 years only last month! That makes UNLV doubly unique, as it turns 50 in the year 2007.

(June 16, 2005)

A CHANGING FACE OF SCHEDULES

It was the summer of 2003. "Satish, why math courses are not offered in a three-week summer session?" inquired the newly hired Department Head. "There is a general belief that math is not popular with students in compressed time. Let us try it next year." I said. Math experiment in the first three-week summer session in 2004 was successful by every count. A new course was added for 2005. Its high enrollment has affected the following two five-week sessions. But overall, the summer sessions have served more students and generated more money for the Department, College, and University.

UNLV has one of the most active summer programs in the country. Nearly 15000 students take courses. Local students studying in out-of-state colleges take transferable courses while enjoying summer at homes. All major universities like USC, Stanford, and John Hopkins accept UNLV's lower division courses. Students save a lot of money!

Generally, faculty believe that math is learnt the way they were taught. However, non-math majors learn it in bits and bites. Also, the business side of education demands that institutions provide as many education delivery systems as possible. It is futuristic.

My thoughts flew off into my academic past. The colleges in India were a different world. At Indiana University, in fall 1968, a section of Modern Algebra course was scheduled in a 3-day format of Tuesdays (T), Thursdays (R) and Saturdays. Soon after, Saturdays were gone. MWF and TR became a norm. At UNLV, in late 1970's, I was surprised to find several courses in education and hotel colleges offered once a week.

The universities situated in cosmopolitan mega cities are confronted with unique challenges. No one wants to burn two hours in driving and parking for one hour of class. Regular driving, in worsening traffic conditions, affects the general health and class performance.

At UNLV, the classroom and office space is not keeping pace with the enrollment growth. On the bases of two surveys of faculty and students, MWF classes are now replaced with MW only.

Faculty and students only come to the campus, if necessary. Under the new 2-day formats of MW and TR, once a week classes on Fridays and Saturdays are also scheduled. Very soon, Sunday classes will be there! Las Vegas is best known in the world as a 24-hour town for quality entertainment, food, shopping, and fitness clubs. So why should it not be known for 7-day education too?

Having taught in both three-week sessions, I must add that students and faculty like it. In my sections of (MATH 095) **Elementary Algebra** and (MATH 132) **Finite Math,** the retention rates were higher and performance better. Immersion technique works in math too! After three weeks, students finish math requirements and faculty collect their pay checks. Being a win-win scenario, math courses in three week sessions are solidly viable.

(June 21, 2005)

WEEKEND MEANS MATH

"There is no weekend, if you take a math course in a summer session. During a three-week session, no party time for two weekends, but have a big bash over the third one. Like wise during a five-week session, forget fun for first four weekends, but have a blast on the fifth one." That is what I bluntly told a student yesterday. He had flatly told me of his not studying over the weekend though the exam was scheduled on Monday. It is largely due to prevalent lax attitude in public schools going on for many years. When these students come back to the college after some experience of the world outside, they really have no study skills or attitude.

The classes that meet from 5-7 PM on Mondays through Thursdays for five weeks provide a greater challenge. There being no class on Fridays, the gap of three days disrupts the study cycle. By the time you go home, eat and rest, it is 10 PM. My rule of thumb for study time is that for every hour in the class you must find at least two hours on the same day at home. But there is hardly one hour left on Thursdays!

Math is unique in many ways. One is the rate at which math evaporates from the top of the head, if not done everyday. There is some inverse square law in effect. For the students, Las Vegas is the most attractive place to earn money. Almost every one holds a good part-time job. Also it is an American way; work through college.

During this 5-week session, 50 % of my students are working full time; it is normal. If a job is typical from 8-5 PM, then where is the study time? It is impossible to squeeze four hours of study before the next class. In fact, 30% of students walk in late! Throughout my professional life, I have been very generous towards lateness and tardiness. As long as a student is genuinely serious in learning, I let them come in late.

From the instructor's vantage point, teaching for four or five days is not very different. The preparation time may not be much, but it is the delivery and management that matter. Grading quizzes and tests before the next class meetings take away a lot of time. Yet it is not as difficult as for the students.

On the first day of 5-week session, I cautioned students about taking two courses with fulltime job. One student missed at least 50 % of classes in an effort to balance job and studies! After three weeks, she dropped one course, and flunking the other. Next session the same story will be repeated! Unless weekends are fully converted into studies, often, a course finishes the student first before a student finishes the course!

(June 24, 2005)

SOME STUDENTS STAND UP

"I am willing to do any thing to pass this course," wrote a student in an e-mail that I read at 11:00 PM yesterday. Immediately, I replied, "Your willingness to do it now, after the session is over, is like buying an insurance policy for covering an accident that has already taken place! On Day # 1, you were told everything about passing the course with specific grades." The course MATH 126 (**Precalculus I**) was over yesterday at 7 PM after the final exam was given.

During the quiet midnight hour, I kept pondering at the growing laisezz-faire attitude of the students. It may not be a reflection of the society as whole, but it is common amongst non-traditional UNLV students. This particular student was repeating the course while working full time. His class standing is junior with major in Mechanical Engineering. But he has not yet finished the first part of precalculus mathematics! I wonder if such students care about academic advising from any one.

The entire five-week session has only nineteen 2-hour class meetings. This student missed five out of last seven classes! It is like missing Week # 11-14 in a 15-week regular semester. In a fast pace summer course, a student can not catch up with the material after skipping even one class. On a casual inquiry about the absences, he told me, without any qualms, that he had gone to North Carolina with his buddies for deep sea fishing! He was proud of his catches and happy about the great time he had there!

This story stretches back to the fourth week. This student was back in town and came to the class on Wednesday. Instead of taking a scheduled test on Thursday at 5 PM, given the end of the fourth week, he came in 30 minutes late, and wanted a make up test! In general, make-up tests are not routinely given, and the policy is explained on Day # 1.

Sometimes, I feel like taking the approach of a doctor or an attorney. As long as clients/patients continue to pay, they are professionally treated. When a student signs up for my course and has the prerequisites, I won't do any thing to turn him/her off from mathematics. That is not equivalent to giving a passing grade at the end. This student earned an F Grade.

(July 08, 2005)

W Means Withdrawal

Last month, the Board of Regents approved a new policy of instituting W grades effective fall 2005. It means that if a student withdraws from a course after the late registration, then W will appear in the transcript. Having no point value, it has no effect on the grade point average (GPA) of the student.

Last week, while talking with a colleague about students quickly dropping courses, it came down to a student who had taken a calculus course at least four times. Also, a course can be repeated to improve the GPA. By and large, UNLV students work for 30-40 hours a week while carrying a heavy course load. A crack is bound to appear in one or both places. If any one has a family, then it would impact a quality of home life too.

This policy is special to me. It was under my chairmanship that last summer the Undergraduate Affairs Council of the College of Sciences recommended W Grade to the Faculty Senate Curriculum Committee. This policy will deter frivolous course withdrawals, phony registration to boost minimum course enrollment, and 'reserve' slots in high demand courses.

On the other hand, it will enhance the retention rates a new buzz word in academe. However, the most important reason is to bring integrity to the transcripts and GPA. **A transcript must show all the academic work attempted**. At the time of admission most professional schools like to differentiate students based on grades; say, one has a B Grade in the first attempt and the other gets a B Grade after two or more attempts. It is like the introduction of +/- grades introduced at UNLV in late 1970's for finer differentiation in students' performance.

It reminds me of fall 1973 at Indiana University. With my math PhD thesis done, I took a graduate course in computer science, *Finite Automata* for enhancing my employability. It was the time of worst job market due to the oil embargo of early 1970's. Being in a different mindset, I neglected to keep up with the course till I saw my sliding performance in a mid term exam. The choice was either to drop the course, or face the consequences. The option of withdrawal was ruled out as never before I had dropped a course. It was partly conditioned by the system of education in India that permitted no dropping of a course. The only C in my transcript is in this course! Nevertheless, I continue to learn from this experience.

These days, a withdrawal from a course is far from being stigmatic. Withdrawal from a course, freely chosen, shows lack of commitment. Suddenly my thoughts flew out over the spectrum. People are no longer committed to any thing! Growing divorce rates, moving away from old neighbors for new houses, or new jobs reflect no allegiance. In the name of enrichment, one could also argue for the other side. A change in routine usually gives a temporary impetus. Such are the signs of our time.

(July 19, 2005)

Run Like A Business, If.......

In the US, whereas, the school teachers are on 12-month contracts, the college and university professors are on 9-month contracts unless one is on a 12-month administrative position. In contrast, the professors in India still enjoy a fully paid summer vacation, and are under no professional obligations.

Academic year in the US runs from mid August through mid May. In the mutual interest of students and faculty, different formats of summer sessions are prevalent in major universities. It provides faculty some compensation while the students are able to finish their degree programs quicker. However, the summer session is not supported by taxpayer money! Instructors are paid according to the money generated by their classes.

I remember spring 1975 and summer approaching when I inquired a colleague in the Business College of his summer plans. He right away said, "I drive trucks coast to coast. It is fun, sight seeing and brings a lot more money than teaching courses." A few faculties are now able to get some research support for summer.

At UNLV, summer session is big in terms of the number of students and the course offerings. In addition to the UNLV students, local students who study out of state, return to Las Vegas, and take summer courses at cheaper rates while staying homes. For the last 25 years, summer school has been a cash cow. A percentage of the summer surplus comes back to the departments for faculty travel and development. **A challenge before the department chairs is to maximize this fund while keeping the faculty satisfied.**

Last year, a newly hired department head said, "For summer sessions, I would rather hire part time instructors!" He was frustrated with regular faculty demanding full pay for their under-enrolled courses, and later on, asking for travel support. His argument was: you get a fair share of what you generate. Students shop for instructors and can easily check their teaching reputation on the web. In the US, everything is ranked and rated!

Here is a moot scenario. Let us say that 20 is the minimum number of students for a full $1500/credit (varies with rank) payment in a typical 5-week summer session. This number is taken from the last Friday before the first day of classes on Monday. The courses are prorated in the sense that one having 15 students gets paid $1175, @ $75/ student. Of course, an instructor can refuse to teach a course. The students dropping the course by the first Friday of classes get 50% refund. In fact, it is very common for 6-8 students to drop a math course for various reasons. That means, whereas the instructor gets fully paid at 20, the summer school goes under red for this course.

After witnessing several scenarios over the years, I suggested that the faculty generating more money at end of the session than they were contracted must be rewarded with a bonus check at the end. **The idea is simple; if it is a business, then run it like a business!**

(August 27, 2005)

PLANTING TREES OF A KIND

Yesterday, while reviewing students' information in the **Honors Seminar** (400H/002), **Topics in Asian Studies,** I was gratified to note that it was meeting several objectives. Some students are taking it to satisfy the Multicultural and/or International requirements in the University's General Education Core. One is majoring in Asian Studies. Of course, everyone is taking it to satisfy the Honors College requirement of four seminars for graduation. These seminars are not open to students outside the Honors College.

This seminar is run by multi faculty drawn from different departments each focusing on one Asian topic for 1-3 weeks. What do I contribute? I focus on non-European roots of science and mathematics in ancient India and China. This semester, other Asian topics are from history, finance, sociology, social work, and pop culture. There is a faculty coordinator assigned by the Honors College.

Also, Asian Studies is an interdisciplinary program housed in the College of Liberal Arts. Being a long standing member of its Advisory Committee, I had proposed this course based upon the expertise of the faculty on Asian topics, or their Asian origins. Sometimes teaching a new course is not as difficult as getting it approved by various committees. Honors College has a fast track for its seminars. Since its unveiling in Fall-2001, it has turned out to be one of the most popular seminars as Fall-2005 runs its 4th edition.

Over the years at UNLV, there are 15 innovative courses that I have designed and taught. Nine of them are one-credit courses scheduled during three-week mini/winter terms existing during 1977-1988. Though they have been taught only once, but each one of them has left a piece of memory behind. Their contents varied from non-traditional topics in mathematics, artificial intelligence and history of mathematics. They are like flower plants that bloom once a year, spread its fragrance, and are gone in a couple of days!

The remaining six are three-credit courses offered during summer sessions at least twice. Three of them, designed for high school teachers, were presented in a teacher academy during 1980's. The other three are seminars in the Honors College. I was the first faculty member from the College of Sciences when the Honors Seminar, **Mathematical Thinking in Liberal Arts** debuted in Fall-1990. The seminar on **Paradoxes in Arts, Science and Mathematics** premiered in Spring-2004.

This *Reflection* has stirred a remark attributed to Prof. John B. Conway of IU/Tennessee (now with the NSF) who once said that in life a man should accomplish three goals: plant a tree, beget a son, and write a book. Yes, I have planted a few surviving trees, though I immensely love cacti! My son has a son now. But, I have not written a book yet! Nevertheless, by planting new courses now and then, I may have uniquely influenced a few lives.

(Aug 31, 2005)

Science Is Beyond The Poor

"**Research is expensive**," said my friend, Bhushan Wadhwa of Cleveland State University. It means researchers, teaching less than those not engaged in research, cost more. Creative thinking requires sustained and long periods of concentration of mind. The 19th century 'prince' of mathematicians, Gauss, called teaching a nuisance in doing research. The movie, *The Beautiful Mind,* based on the life of the Nobel Laureate, mathematical economist, John Nash, captures this scenario. When he walks out to teach, he leaves his mind behind in the office where it was engaged in a research problem!

Last Tuesday, I gave a three-hour lecture in an interdisciplinary Honors Seminar. Nearly three fourths of the students have taken only 2-3 general courses in science and math. It was essential to devote an hour on the organized nature of these disciplines. Though science and math have pyramidical structures, but research methodologies are very different. Math, by and large, is a loner intellectual sport. But science, being experimental, involves a team of at least a few graduate students and laboratory.

In order to put it in perspective for the students, I asked, "How much salary a new PhD in history gets at UNLV?" Quite a few guessed it around $40,000. When the same question was raised for the sciences, they were still around 40 K! I had to remind them that they were not living in a socialist country, but in the US where free supply and demand determine the price. A starting salary in a science area for a raw PhD is around $60,000.

As UNLV heads towards research, a new dimension of start-up fund is added for the new hires. It is the money given to the new faculties to set up their research labs. The average tab is $250,000/new hire. Until I became associate dean, I had no idea of this expense. That is one instance, how research in sciences is relatively expensive. Many small and underdeveloped countries will remain light years behind the US in fundamental research.

Nevertheless, a pertinent question remains: **Is science relatively more useful for the society?** The university administrators, regents, and law makers have to tackle it periodically. The year 2005 is the centennial of Einstein's miracle year when he published four research papers that changed the course of mankind. Talking of the impact, there is hardly a day when a play of Shakespeare is not enacted in some part of the world. In contrast, there are no public celebrations of Einstein's work! **The very nature of science and math makes any great discovery diminish with time.**

With the start-up funds, the faculties come under pressure to go after external research grants. If the start-up money is 'recovered' in 5-6 years, then it was well 'invested'. It is time to encourage science in nodal areas so that 2-3 faculty members nurture their research in the same lab. Otherwise, on retirement, or resignation of a faculty, his/her entire lab may have 'junk' value to other researchers. This reminds me of a famous quote of the 1960's - **The American economy thrives on waste!**

(Sep 08, 2005)

TIME TO STOP ERASING

Yesterday, it was kind of amusing to watch a few students hurriedly erasing their solutions from the worksheets and overwriting them during a 7-minute pop quiz. I stress, "During a test or quiz, time is as precious as water during a desert hike. Therefore, never waste time in erasing a solution, or 'whiting' it out. If you are sure of what has been done is wrong, then put a gentle cross on it, and move on to a clean workspace."

The American students are conditioned to work on problems in marked spaces, or stick to the ones they start on. It is like occupying the same seats in the class that they take on Day 1. On the top of wasting time in erasing, in a majority cases, the second solution is often incorrect! Sometimes, the erased effort is better.

The Yellow Pencil Number 2 has been an American symbol for more than 100 years. No one can ban its use! However, for the last many years, I have been encouraging the use of ink pens during tests and quizzes by giving an extra point! Is that a discrimination against those who use pencils on the tests? No; students don't object to it as it is explained early in a semester. There is no penalty for the pencil users.

My reasons are more than saving precious seconds. "Unless you zoom in at your mistakes they are going to come back. Once erased, you don't see them. Test time is a stressful." Extending this thought to life in general, 80% of our mistakes are old! Unless one looks into the eye of a mistake, it becomes dormant and jumps out like a virus, when the immune system is weak. A preparation for a test is to go over the old mistakes.

My grading is not binary. One gets zero only for not attempting a problem. A mere answer is not a solution; complete work alone gets full credits. To make students understand, I would once-in-a-while grade a crossed out problem and give partial credits. That makes an impression on the importance of writing in mathematics.

I don't know any history of the pencil usage in the US schools and colleges. But recent trends of exams, like using scantrons and blackening the ovals, may have encouraged it. During my hoary student days in India, we had two sets of study routines. In class, notes were taken with pencils on scratch notebooks. At home, the class notes were redone in ink on the so-called, fair notebooks. It is now funny to recall the inkpots, penholders, and various nibs used till my high school. Fountain pens were for the college. It was 1950's; owning a Parker fountain pen was like driving a Mercedes today!

The ink pens also help in academic integrity. After the grading, the quizzes and tests are usually returned to the students. It is then very tempting on the part of a student to erase a wrong solution, done with a pencil, and re-do it correctly, and then tell the instructor at the end of the period that a correct solution was wrongfully graded! The ink writing is not easily erasable after a few hours. Who knows, erasable ink may also be in the market.

(Sep 08, 2005)

A Trial Of Belief Systems

Writing a sequel to my grandson, Anex's hospitalization (Taj's *Reflection* name) at the age of four weeks, has been on my mind for a long time. Last week, a question cropped up on the make-up of a scientific mind. It was during dialogues with the students in an Honors Seminar: *Topics in Asian Studies*. That is how Anex became a part of my segment of lectures in this multi-faculty seminar. The discussion continued over two weekly meetings (Sept. 06 and 13), each lasting for nearly three hours.

Life's deeper beliefs are tested during a period of crisis. Anex was born three weeks earlier, and he spent exactly three weeks in the pediatrics intensive care unit of the same Catholic Hospital! In a doctor's office, as soon as he was checked for high fever, the doctor called the ambulance, hospital emergency, and gave instructions to the physician-in-charge at the emergency about Anex's grave and deteriorating condition.

If this were delayed by a few minutes, the meningitis may have proved fatal. For nearly 72 hours the little 7 Lbs lump of flesh was all wired up with various machines and computers monitoring his every breath. A nurse posted outside his room had a special monitor. The parents, relatives and friends are helpless spectators in modern high-tech medical care system. It is in contrast to the old fashioned medical systems in India when every family member and friend would run around for various chores and errands between home and hospital, and any stop in between.

As the word of Anex's critical condition spread, there were sympathetic calls and suggestions for supplementary and complementary remedies. A well-meaning friend wanted a huge picture of Puttapurthy Sai Baba's prominently placed at the head of Anex's crib. When asked to bring a 5"x7", he essentially withdrew. It reminded me of advertising banners prominently placed in the corners of a boxing ring for the TV viewers. My wife had Lord Ganesha's idol placed in the hospital room. A nephew-in-law, whose son had undergone a long hospitalization at the time of birth, suggested some holy water. This water is collected after a ritual bathing of an idol or a picture of a god man. My wife, sister, and brother-in-law known for his analytical mind, recited a special healing *MAHAMANTRA* for several weeks.

All this and more was happening while the tiny life was hanging by a thread. My daughter was fraught with pain and felt unforgiving towards herself. Every hour was an eternity. My sister called her close friend in Delhi who called her parents in Bathinda, and they consulted a family astrologer in Faridabad. A special *POOJA* (prayer+rituals) and its exact timings were relayed back. With necessary ingredients, the *Pooja* was meticulously performed partly at our home and partly in Anex's hospital room.

During this period, our little soul went through every diagnostic test, including MRIs and ultrasounds. Once after several insertions, when the nurses could not find a vein in his tiny arms and legs, they went through his skull! Eventually, the bacterial and viral causes were nailed down and flushed out of his system one by one.

I witnessed the whole panorama. The hospital nurses, physicians and administrators took care of Anex in a remarkable manner. They explained every action and outcome at each stage. They never held back any information from my daughter and son-in-law. Every question was answered to their satisfaction. At the same time, friends and relatives on both sides of Anex were doing their best according their belief systems. Academically speaking, did we forsake a proven scientific system over an unscientific one? That is the question I posed before my students.

I am convinced that at one time, there was a **Science of Life Force (*PRAAN SHAKTI*), and a paper was presented six years ago. Like many great treasures of Indian heritage, it is lost during a thousand years of foreign subjugation of India.** Its scattered traces are seen now and then. I couldn't be a passive spectator under the circumstances. Every day, during moments of my dis-engagement from the official work and tranquil hours at home, I harnessed my mental energy and telepathically passed onto Anex.

After a week, I stopped one early morning at the hospital before going out of town. Feeling fully charged, I entered his room, washed my hands and covered his tiny head with my right hand and mentally tried to suck all the 'negativity' from his system. After a few minutes, I symbolically gathered the 'negativity' in my hand and threw it in a trash can. After washing the hands again, I spoke in Anex's ears, "When I come back tomorrow, I shall see you in a different room (for patients with downgraded conditions)."

Analyzing my actions and pondering over them: Did I really forsake my faith in the ultra scientific treatment? Unqualified, no. Do I give any credit to all the rest that went around Anex? The answer is a qualified no; because cause and effect are not clearly connected.

Life thrives on faith transcending from rational thinking. Conversely, rationality also leaps out from blind faith! One just cannot argue over them during moments of life and death. The prayers were held in social gatherings. Our son-in-law, being a Christian and having relatives in Ecuador and Mexico, had special church prayers held in the US and overseas. Prayers and best wishes are common in every culture. **It is a sign of man reaching out to the Ultimate Limit for the realization of a far reaching goal, when he falls short in human efforts for the time being**.

In Indian folklore, there are numerous lores and legends when a curse works with efficacy as does a blessing. That being a case, then at least 10% of 1.6 billion Muslims cursing America during Friday prayers all over the world must have an impact! The US must take its security very seriously. Being manifestation of the same Life Force, blessing and curse are two sides of the same coin. **That is how the story of Anex gets tied up with a developed mind based on the modern science of the West and the 'lost' yogic science of India!**

(Sep 15, 2005)

BIRTHING PAINS OF PROOF

Last week, it was a bit shocking to notice that not even a single student got a particular problem right in a class test. It was proving a statement that was discussed in the class! In order to impress that the only way to understand a math problem 100% is re-doing it by hand. My tests contain at least one problem that is already discussed. Barring a few exceptions, one initially learns any skill by watching a master doing it, and then copying it. There is a template for any activity. It is specifically instructed that all the proofs done in the class must be re-done as a part of homework.

The essence of mathematics lies in proving statements. Generally, a course on discrete mathematics is the first place where the US students are immersed in this aspect of mathematics. In the lower division courses, proofs are minimally done, or just omitted. Most students in this course are sophomores and juniors majoring in mathematics or computer science. With a majority of students, there is a genuine struggle on proofs.

It reminds me of a 1959-story when I was doing my master's. The math curriculum at the undergraduate and graduate levels in India during that era was essentially of mathematical physics by the present US standards. In a real analysis course, based on then unheard point set topology, a classmate was driven over the edge of frustration. Every time he tried to reproduce a proof done in the class, he could not get it. It happened not once or twice, but 18 times! He was not an ordinary kid. He was ranked 23 in the state amongst nearly 45,000 students taking exams for their bachelor's degrees.

My first exposure to proof was through classical Euclidean Geometry in the 10th grade. Algebra was mastered in the 9th grade. But the early frustration over proof is still vivid in my memory, though it happened in 1954! After two weeks of suffering, the older brother of a classmate explained it in a manner that suddenly I felt as if a bulb of understanding had lighted up inside my head. Eventually, I enjoyed the deductive reasoning in proofs in Euclidean Geometry far more than the trickstery in algebra.

I believe that Euclidean Geometry should replace the axiomatic geometry in high schools. Nothing in the entire gamut of mathematics is as elementary and elegant as different sets of propositions spread over thirteen books comprising the *Elements* written by Euclid more than 2300 years ago. No where else, one can get clearer ideas on postulates, axioms, definitions and proofs. I offered some material in a one credit experimental course, *Back to Euclid* during a 3-week winter mini term in 1988.

Most US high school teachers and students dislike the current contents of geometry. As a parent, I also anguished when my kids went through it. The inclusion of Euclidean Geometry in high schools will make the college discrete math course, as introduced in 1980's, redundant. The proofs done in regular math courses will be comprehensible to students.

(Nov 05, 2005)

TEACHING AWAKENS CREATIVITY

It was gratifying to read an article, *Faculty should Stress Creativity on Campus* published in the *Rebel Yell* (11/28/05). The writer, being an undergraduate student, made me happier. Students demanding creativity in instruction is a good sign for UNLV in order for it to mature into a first rate institution.

The article compelled me to examine creativity in my life as student and faculty member. During four years (1955-59) of college in India, one math professor, apart from other shortcomings, never covered the syllabus. The final comprehensive exams, given at the end of the year, were written by the external examiners. The course instructors had no clues of the exam problems. To top it, the final exam carried the entire grade! It equally tested the instructors as to how well they had prepared their students.

In order to make up for that professor's instruction, we used to borrow notes, seek help books, and form study groups. I really worked hard on his material. It paid off later at Indiana University while working for my doctoral thesis. Before I grasped a research problem, my thesis supervisor had to go overseas for a year. In his absence, I diligently and independently finished the research work enough for a thesis. The professor whom I criticized for years, I acknowledged him in reverse gratitude for developing my independent thinking and work habits! **Independent mind is a creative mind**.

The *Rebel Yell* article comes out strong against faculty using power points and transparencies. That is not teaching, but regurgitating. However, a colleague with health problems could never teach without this technology. Teaching math courses using transparencies is like explaining worked examples in the textbook. **Without student participation, a spark of creativity can not be ignited**. The essence of mathematics lies in deductive reasoning.

Getting lost, while discussing a problem, or stumped by a student's question, is never to be taken as an embarrassing moment. As a student, I learnt more from those rare moments when an instructor unsuccessfully tried one approach after the other. However, every instructor knows that not every student benefits from such a predicament.

Creativity, unlike pancakes and sausages, is not seen every day. There are hours of no activity, frustration and hopelessness before a silver lining of inventiveness breaks through the fog. Originality cracks open only after the mind remains agitated for a long time. Teaching, structured in a dialog form, lets the students experience ingenuity.

Often the faculty inadvertently stifles the creativity of the students. A colleague would give no points, if certain steps, in a solution, were not done in the manner he wanted. If the answer to a problem was not written in its answer box, he would mark it wrong. **Students may thus learn some techniques from him, but at the cost of their smothered creativity**.

It was during my master's (1959-61) that I had a taste of creative thinking in mathematics due to the participatory and intimate style of Professor H. R. Gupta. I had him for nearly three semesters. **Creativity is never taught; only skills are taught.** An instructor can provide incubation for creativity with his/her **competence** over the subject, **compassion** for the students, and **communication** of matter. These three C's are the pillars of my teaching philosophy.

The vitality in classroom is maintained, over the years, by teaching different courses and changing textbooks. I have taught a variety of 46 different regular courses at every level. In addition, 15 new courses were designed and offered as experimentals and seminars. Innovative courses usually draw innovative students. Currently, I am working on the 16th course. Creativity is contagious and rubs off, if a student takes 2-3 courses.

Creativity, being not exclusive to a discipline, I encourage my students to write 400-500 word reports on relating math with their majors, favorite hobbies, campus lectures, popular movies, books, etc.

Students have written some of the most interesting short papers. The very thought of making bridges over the farthest points is challenging. A conceptual connection, between any two realms, is creativity. It can be integrated in any course.

A week ago, a non-traditional graduate student asked,"Dr. Bhatnagar, what interesting courses are you teaching next spring?" "I can make any course interesting now!" was my immediate response, may be a bit stretched. However, it defines my professional life.

Generally, the students have been very objective in their evaluation of my teaching. Over the years, I have given away hundreds of books, research papers, and class notes to inspire the students. At the same time, unloading of books keeps in check the shelf space in my office. The only thing that are still there are boxes containing all student evaluation sheets.

Talking of student evaluation sheets, within a few years, they will be history in a junk pile. The website, **RateMyProfessors.com** is doing a fantastic job on the evaluation of teaching. It is a great resource for students to checkout instructors who are easy, creative etc. From this site, faculty can find trends in instruction from students' perspective. Like the library citation index for judging research, three teaching scores used in this website, may be considered for promotion, tenure and merit, in years to come.

Teaching at UNLV can be negotiated for research and service. Presently, the new faculty in the College of Sciences teaches one course or none, during the first year. However, I taught eight different courses during my first year (1974-75). That was a different era. Borrowing a refrain from Albert Einstein on Gandhi, the coming generations of faculty will find it difficult to believe, that there lived a professor who taught over 60 different courses at UNLV!

(Dec 03, 2005)

Two Tales Of Mathematics

During Dec 26-29, 2005, the 71ˢᵗ annual conference of the Indian Mathematical Society (IMS) was held in Rorkee, India. Last week (Jan 11-15, 2006), it was the Joint Mathematics Meetings; the 108ᵗʰ meeting of the American Mathematical Society (AMS), 85ᵗʰ of the Mathematical Association of America (MAA), and other small mathematics organizations, held in San Antonio, Texas. It is natural to reflect on this rare double experience never had before.

I became a life member of the IMS during my two year sojourn (1980-82) in India. Except for the first year, never any communication has been received from the Society! The IMS has no permanent office like the AMS and MAA have them. Its office charges physically move with the election of new officers. **Though the Indians today are world famous as IT experts, but the IMS maintains no website!**

Like many learned societies in India, the IMS was founded in 1907 by the British scholars coming on various assignments to India. Gradually, Indians took it over. It is interesting to note that the AMS, founded in 1888, has over 28,000 members and the MAA, in 1920, has 30,000. I switched my membership from the AMS to MAA in 1976.

The social clubs and professional organizations are the hallmarks of the organized western mind. The Hindus being highly individualistic, all their organizations are seen to flounder eventually. There are varied annual membership dues for the AMS and MAA, but none for the IMS except its lifetime membership. It is sheer paradox, that whereas the number of students doing master's and PhDs in math, and of professors teaching math in colleges and universities is perhaps four times that of the US, yet the IMS has hardly grown in 50 years. The President of the IMS told me that the attendance at the annual meetings has been around 100!

My first memory of an IMS meeting goes back to 1960 when I attended it in Chandigarh as a student (volunteer) doing MA (Part

II). My second meeting was in 1986 while on sabbatical leave in India. It was held in Jaipur, and presided by my teacher and mentor, Professor S D Chopra. The Roorkee Conference was presided by Sarva Jit Singh, known to me since 1965. Both of us were PhD students of Professor S D Chopra in Kurukshetra University (KU). I left KU in 1967 without finishing the thesis.

It was heartening to watch Sarva Jit give the presidential remarks followed by his technical address. One feels honored in watching a friend honored! The entire gathering was accommodated in the auditorium of the mathematics and physics departments of IIT Roorkee. As a historical footnote, the present IIT Roorkee was started in 1847 as the first engineering college in the British Empire. Yes, the entire conference deliberations, paper sessions, and stay of the delegates were on the campus.

In contrast, the San Antonio Meetings was held in the San Antonio Convention Hall. It was attended by over 5000 mathematicians, students, spouses from all over the US and the world. It is the greatest spectacle of mathematics on earth! There was a Press Room for daily releases. Thousands of papers were presented in hundreds of sessions in addition to various committee meetings, workshops, exhibits, and publishers. The high-tech Employment Centre alone covered an area of 300,000 sq ft. It draws employers and job seekers from all over the world.

The mission of the Joint Meetings is to promote mathematics at every forum with teaching innovations, integrating technology, curricular reforms, publications, and latest research in every conceivable area of mathematics. My favorite sessions are ***Sports and Mathematics***, ***Arts and Mathematics*** and ***History of Mathematics***. There were 4-5 booths in the Exhibits promoting mathematical concepts printed on T-shirts, hewn in stones, chiseled in wood, and designed in jewelry. One can soak mathematics of whatever taste and style one likes. Eyes cannot see it all, and mind cannot comprehend it either!

The delegates, staying in about ten hotels, boost the local economy. They are welcomed by every establishment. It is all a business and the Americans are best at it. Each mathematics organization has a fulltime Executive Director and staff who run it with the policies laid out by its elected Board. Members pay dues and in return get the services and benefits of a professional organization.

In contrast, the IMS seem to be spinning around a couple of individuals who are prominent by special VIP badges on their lapels and a cluster of individuals around them. It is a typical Indian style where an individual in office works to enhance his/her image and not build the organization. The western mind generally leaves the office stronger than before. It defines the functioning styles of east and west.

One wonders as to how the expenses of the IMS conferences are borne when there is little source of regular revenue. Since the entire show involves about 100 people the expenses are relatively small too. The host institution bears the bulk. Being vacation time on the campus, the classrooms, auditoriums, dorms, and guest houses for the VIPs are freely available for the conference activities. Also, some state and central government agencies pitch in financial support.

The highlights of the IMS are the refreshments during tea/coffee breaks, luncheons and dinners included in the registration. It provides an intimate atmosphere for interaction. To top it are free cultural programs in the evenings by the local artists. That involves a tremendous organizational work on the part of the host institution. In the US, the local institutions may have some advisory role, but the organization of the conference is mostly done professionally. Such a great event has a template that the core staff uses every year, thus improving and adjusting it according to local conditions.

I attend these conventions for a couple of reasons. Number one is social, in meeting old friends and making new ones. Besides, presenting a paper, or organizing a session, as I did this year, on **Philosophy of Mathematics**, I love to attend various invited addressees for getting global views on mathematical issues, prize functions to applaud the new stars of mathematics, and technical papers to keep my mathematical muscles in shape. Any thing else is bonus! At one time, I used to stay for all the four days, but now two days suffice.

It is a universal human urge to see and meet a legend in one's field. A great American mathematician, P. R. Halmos (I took a course from him) once remarked, "One must support a math organization by regular attendance." Halmos has not attended it for the last couple of years for health reasons at 90, but I did spot 80 year old agile and silvery Peter Lax, the 2005 winner of the Abel Prize (equivalent to the Nobel Prize in money). After finding a seat in the networking area, I enjoy writing and observing people associated with mathematics.

The aspect of obligatory attendance of the conference was absent in the IMS. A couple of my former teachers and friends who had served as presidents of the IMS did not want to travel a distance of mere 100 miles to attend the conference. **At the IMS meeting, either you are somebody as an invited speaker, or nobody.** Can this culture change? It is an open question that few Indian mathematicians would like to tackle.

For many years after settling in US, I did not appreciate the power and benefits of professional organizations. If the US is leading in its world class math departments, PhD programs, innovative courses, and cutting edge research, then it is because of the people working together through the AMS, MAA, and other professional organizations.

Incidentally, P.R. Halmos and his wife donated nearly 3 million dollars to the MAA a couple of years ago. Voluntary donation to professional organizations beside regular dues further sets the west apart. The Indian mathematicians have a long way to go in giving donations to promote their own discipline.

The US math convention is like a mini UN as one can see people from every ethnic group and nationality. I saw relatively more Indians in the San Antonio convention than at the Roorkee conference! At the IMS meeting, only two Sikhs and one Muslim were there. There was no way to identify Christians.

On noticing the Muslim alone, I introduced myself. He was from Iran and doing PhD in Pune! That speaks of the caliber of individual Indian professors and some universities to attract graduate students from overseas. As a sidebar observation, there is a statistical correlation between the religious beliefs and the study of abstracts areas of mathematics.

(Jan 16, 2006)

WHO IS A GREAT MATHEMATICIAN?

This question is an offshoot of my recent *Mathematical Reflection* in which I mentioned P. R. Halmos as a great American mathematician. A few comments were received, that in turn re-set my mind to contemplate over it. After all, I have already spent 45 years in the service of mathematics. Question: How to measure the greatness of a mathematician?

Last night, I witnessed Kobe Bryant scoring 81 points in a game. A commentator described him a basketball machine. The basketball fans will remember him a great shooter. In every sport, the players are complete in more than one aspect of a game. Nevertheless, an individual performance comes next to the team performance.

That is not the general perception of a great mathematician. The modern history of mathematics is mostly Eurocentric. The great Hindu and Chinese mathematicians of the past are isolated in the sense that little is known of their contributions, and schools of thought. Their long colorizations have made it worse. Until, the middle of the 19th century, most mathematicians were essentially mathematical physicists, or applied mathematician in the lexicon of 20th century mathematics.

Taking recent history as guide, the world remembers those who have enhanced the progress of mathematics by proving new theorems, or solving great problems. Ironically, the world knows nothing about some solid mathematicians who spent all their lives on very difficult and unsolved problems like Riemann Hypothesis or Goldbach Conjecture. Also, no matter how inspiring math teachers have been, history does not give even footnotes to them. That is a global view of professionals in mathematics today.

Another thought was to compare mathematics to electrical power. Electrical engineers are involved in three stages; power generation, transmission, and distribution. By and large, mathematics, too, has three components: generating new math by proving new theorems, transmitting math by teaching and cultivating new generations, distributing mathematical knowledge by textbook publications, or applying it in science and technology.

The question of the greatest mathematician is not like the highest peak of a mountain range. In the US today, men and women, schools and colleges, teams and institutes are professionally ranked everyday. Mathematics departments are ranked too. Though there is no Nobel Prize in mathematics, but Fields Medal, Abel Prize do provide a measure of mathematical research. The annual teaching and service awards of the MAA and AMS have put national spot light away from research. But they are limited to the US only. Teaching is so cultural dependent, that the comparisons are difficult across nations.

In the US alone, mathematics is taught in over 5000 colleges and universities. Who is overall a great mathematician of a year/decade in terms of research, teaching and service? It is a good question. Beyond its discussion over drinks, it answer will remain 'local'.

(Jan 2006/Sep 2008)

John Green: A Purveyor Of Mathematics

Students at the commuting campuses in the US metropolitan cities have similar characteristics, but the Las Vegas touch makes UNLV students really unique. After teaching here since 1974, and seeing national surveys of student bodies, UNLV is in its own league. It is fun and challenge to face these working, matured and secured students!

John Green is a fixture not only in Mathematics Dept, but also at UNLV. He is 60 years old, 6'- 8" tall, and weighing over 220 lbs. These physical statistics match with his intellectual 'statistics' for his passion for teaching mathematics, breadth in mathematical knowledge, and remaining excited about doing math problems. I have known him since I joined UNLV. He also took one upper division course from me.

John is a Vietnam War Veteran and its streak is seen in every aspect of his life. He loves to take all kinds of courses but has no patience to work on a thesis! Consequently, he did not finish his MS in Physics, Math, and an MBA! Yes, he has enough graduate credits for three MS degrees, if course options were there in master's programs!

John loves to engage anyone into his teaching including math problems he assigns students in homework, tests, and student performance and final grades. A few years ago, he took my copy of Loni's classic on trigonometry having most difficult problems on the subject. He used it to challenge his best students. He himself had big laughs while working them out. Most faculty members stay away from teaching Precalculus II since it includes trigonometry and other topics. In fact, one colleague retired teaching only Precalculus I every semester, but never precalculus II!

John's services to the Dept are invaluable. As a part time instructor (PTI) teaching 1-4 courses per semester, UNLV must have made a small fortune off his teaching. Whenever, he is not assigned any course, he doubles up on his practice of private tutoring. He is one of the most sought after math tutors, provided one can withstand his booming voice and fringe life style. The students, tutored by him, significantly improve their class performance, and others taught by him are known to do better in their next math courses.

A month ago, John met a very serious accident that put him in a hospital for a week and now in rehabilitation for a while. Last week, when the university announcement for the best PTI awards came out, I thought it was time to nominate John and see him recognized. Math Dept has some of the finest PTIs and some even have PhDs. But John's record is the longest and strongest!

Being an exponent of a course option for most MS degrees, if I were empowered to confer honorary master's degrees, then John Green would have not one, but three of them in one shot!

(Feb. 04, 2006)

Michael Golberg: A Pure Researcher

"Do you know what he was discussing with me?" asked my colleague in the hallways this morning. I said, "I know it." "Research problems!", he prompted. We were talking about Michael Golberg (not Goldberg!) whose left leg was amputated three days ago. Fortunately, the right leg is saved for the time being. The legs were infected with gangrene caused by rampaging diabetes. Parting away, I added that **Michael lost his leg not to this disease, but to his passion for research**. He totally neglected his health.

With my 45-year experience as a math student and professional in India and USA, if I were to identify 2-3 top researchers in mathematics that I have personally known, then Michael is one of them. He came to UNLV from MIT with ABD (All But Dissertation) in 1968. The visiting position that I joined at UNLV in 1974 fall was vacant due to Michael being gone on sabbatical leave.

Imagine the academic conditions at UNLV, a Carnegie Master I institution in 1960's. Teaching 3-4 courses per semester was normal. Research was not even in the air! But Michael was possessed with it. It all came from within, a self motivator, and beyond the specter of carrot or stick. He was a lone fountain of research in the desert. His incredible quality is to engage any one, with any math background, into his research problems!

Michael went to MIT for PhD after his bachelor's from Montreal, but left it after strong differences with his supervisor. It is the combustible fuel of anger and frustration that he ploughed back into his researches. A steady stream of quality research papers by an ABD embarrassed some MIT faculty. A story goes that MIT contacted Michael to submit his published research as PhD thesis that he refused. But his wife put his thesis together! Michael never went to MIT again; the diploma came in the mail.

A saying, that a top researcher is also a top teacher; I haven't seen it in my life so far. How good Michael was as a teacher? He was always a researcher! A couple of his students, who went on to do PhD in mathematics, told that he covered his on-hand research in every course. He taught only a few upper division and graduate students, and showered them with A grades. Perhaps that was justified by the challenging material presented.

Michael considered conferences as a waste of time, and avoided airlines. He told me that once a mathematics theorem was proved in aerodynamics, then he would travel by air! **But Michael really helped fly the careers of a dozen of persons**. He is the author of nearly 200 papers, and 5-6 textbooks, research monographs and proceedings. His co-authors are from an algebraist to a statistician, and any specialization in between.

A colleague once told me that despite ill health Michael can keep 10 mathematicians busy with his research ideas. In 1990, he took early retirement due to medical reasons. He is perhaps the only free lance research mathematician in Nevada!

(Feb 08, 2006)

Happy 70th Birthday, Paul Aizley!!

I don't think any one can claim to know Paul very well except perhaps, his soul mate, wife! He has the persona of an administrator, but also loves to teach. As dean, he always taught a course. It was a Calculus III that Paul team taught with a physics professor that my son, Avnish took it, in 1986. While the two professors often argued over mathematical rigor vs. pragmatic approach, the students thought they were confused. Yes, it did confuse some students. However, Avnish grasped different approaches to the extent, that he switched his major from Computer Engineering to Mathematics!

To the best of my backtrack reading, young Paul had an ambition to become a university president?! This highway generally goes through the offices of department chair and college dean. Realizing it was not easy to out-vote the current chairman, Paul tried a short cut for gaining the administrative experience as Deputy to President Baepler. In fact, his moving up in administration coincided with my joining UNLV, in 1974!

When President Baepler moved up as the Chancellor of the State University System, Paul came down to the realities of faculty in 1979. By then, he was well established in the university system, and had chaired the Faculty Senate twice! Despite his exceptional service and teaching record, full professorship was a battle. He directly took his case to the Regents, and won it over the consternation of the then President Goodall.

That is perhaps the beginning of Paul's run-ins with the presidents. When in 1984 Maxson took over the reins of UNLV, with chairman still entrenched, Paul went back into administration as summer school director, and gradually expanded his empire. He was all set to retire as dean, but found President Harter too difficult to work with.

Paul made no bones about his displeasure over the Harter Administration. Had he won the Regent election, Harter would have been history a year earlier. But another tough lady Thalia Dondero stopped Paul into the tracks. Watch for such ladies in your retirement!

A person, new on Campus, wondered at Paul being elected to the Faculty Senate. I told him that Paul's name was a fixture at UNLV. In fact, Paul is the first faculty member to win a Senate At-Large seat normally won by Pro-staff.

Paul's third shot to become the Senate Chair did not work. He misjudged the forces for and against him. With Jim Rogers as Chancellor it was a matter of time before Harter was gone. Feeling vindicated, Paul, you may help Rogers in search for a new president!

Paul returned to the Dept in 2003, but suddenly decided to retire in phase. Paul is witty and sharp with one liners. He carries an enigmatic smile of a male Mona Lisa, if there is one! With such a health body and mind, Paul, you have many years of productivity and service to the community. It has been my pleasure to know you. All the best of the Day!

(Feb 11, 2006)

An Euler I Know

Life is defined by unforgettable experiences. The greater an event, the longer is its delayed reaction before it registers on the mind. One such episode took place last Sunday when I visited Michael Golberg (not Goldberg) in hospital. My visit being unanticipated, he was a bit surprised. He was on the phone and his wife sitting on a chair set parallel to his bed. There being no hurry on my part, I motioned him to finish the conversation. I continued to reflect on Michael as I have known him since 1974. He retired in 1990 for medical reasons. He was only 50 then.

It did not take me any time to figure out that he was discussing a math problem with one of his research collaborators. Michael's left leg was amputated three weeks ago. The right leg escaped amputation due to an advanced bypass surgery done on it by a reputed heart surgeon who had also performed a multiple bypass surgery on Michael's heart a few years ago. The diabetes has ravaged his body. It had affected his kidneys to the extent that twice a week he is put on dialysis. His right eye is totally lost, and he hardly sees with the other; had eye surgeries too. Why did Michael suffer from multiple health problems? One guess is diabetes; caused by sedentary routines, and that was due to his life time passion for research. It is cyclic. There are, of course, other factors too.

After 15 minutes, his wife asked Michael to stop and switch conversation over with me. **Michael cannot disengage his mind from mathematics**. He is consumed with mathematics and won't stay on any other topic. However, when I read him my earlier *Reflection* on him, his eyes were moistened. Soon he occupied me in his researches. He told me about a research problem that his research group was the first to crack. While he could not continue on it for some reasons, another research group in England extended his work. He felt it a kind of loss! Michael is very competitive and stays at the cutting edges in his research areas.

Remembering my once-upon-a-time research interest in partial differential equations, he explained me of a boundary value problem involving Laplace Equation. In 1920's, a well known French mathematician, Hadamard (1865-1963) had categorized it as an ill posed problem, hence not much research has been done in that direction. Michael claims a handle on this long standing problem by combining the analytic methods with techniques of numerical analysis and statistical inference. His current collaborators include experts in all three areas.

"Michael, what continues to drive you for research?" I pointedly asked him. Michael was very honest and forthright in his response. He recalled a long phase of mental depression when he had nearly quit on life. He told me a story, and credited his recovery to a mathematician relative and his present wife. When I pushed him further on his ability to do fundamental research, I thought he would say that he wanted to transcend his body pains and sufferings. But what he spoke, it just took my breath away. He said, **"If it were not my health problems, I would not have done this body of research!"**

There was no trace of any pain or aguish on his face. It is no reconciliation with his conditions, but equanimity of mind. Michael is a living example of mind over matter, the undaunted human spirit, self-realization, the ***GIAN YOGA*** (The Path of Knowledge) of the Hindus, or the ***TAO*** of the Chinese!

Michael is an Euler on some scale in my life time. Leonard Euler (1707-1783) produced more than half of his research papers after going blind during the last 10-12 years of his life. Michael also reminds me of the great Indian mathematician, Sriniwas Ramanujan (1887-1920) who did epochal research from a Cambridge Hospital. He was admitted for unabated fever caused by undiagnosed tuberculosis rampant in England during its period of early industrialization.

Michael does not belong to an academic institution any more. He is a free lance researcher. But he is truly a one-man-research institution! I think the American Mathematical Society, the Mathematical Association of America and the Society of Industrial and Applied Mathematics should come forward with academic support and special recognition for Michael Golberg.

(Feb 17, 2006)

HOLOCAUST AND GODEL'S THEOREM

What a powerful day was it today! It coincided with **Baisakhi,** the beginning of a New Year in India. One half of it was for a lecture scheduled in the afternoon to commemorate the birth centenary of Kurt Gödel (1906-1978) who in 1931 proved one of the most famous theorems in entire mathematics. Ivor Grattan-Guinness, Emeritus Professor of History of Mathematics and Logic (Middlesex University, UK) spoke on *The Reception of Gödel's Incompeletability Theorems by Logicians and Mathematicians, 1931-1960.*

Guinness detailed how this deep theorem, specifically in the area of mathematical logic, was not given its dues by contemporary logicians and generations of mathematicians till 1950's. **It was an 'intellectual' holocaust of the Theorem!** The world of mathematics was dominated by the German mathematician, David Hilbert (1862-1943) whose celebrated plan (1920) to formalize mathematics was smashed by Gödel's Theorem. Guinness described Hilbert's influence on mathematics as that of a Field Marshal! Hilbert witnessed the purge of Jewish mathematicians at Gottingen. Gödel's Jewish hereditary is uncertain, but he escaped to USA in 1940 via Russia and Japan.

The other half of the day, I walked in unknowingly. After a couple of hours working on my office PC, I often step out to straighten my neck, back and hands. The UNLV's Barrick Museum of Natural History is a place to check out for new exhibits. Today, on display were 85 black and white photographs. I often observe a collage of paintings or photographs without looking at the names of the artists, titles, or other description.

It did not take me more than 6-7 frames and a few minutes to recognize them as the shots of Nazis concentration camps, though there was not even a single human face in any one of them! Nevertheless, my mind provided the missing images of 11 millions people systematically exterminated like cattle processed in modern packing plants today.

The pictures, taken from various angles of the camps, are relatively recent and taken by the photo artist, Michael Kenna during 1988-2000. The collection is called, *Impossible to Forget*. The photo exhibition and an hour long documentary, *Memory of the Camps* (Frontline PBS, 1985) are timed with April 1945 liberation of concentration camps after the Nazi surrender.

The documentary has only April 1945 photos and sound tracks recorded within two weeks of the Allied Forces' take over of more than 300 concentration camps in Germany and its occupied countries. **For reasons unclear, it was after 40 years this material was turned into a documentary**! I thought I had read and seen everything that went inside these camps. But here the scenes are unworldly enough to shake one's faith in humanity.

A human paradox is also presented there. The people living in cities only a couple of miles away from the concentration camps either had no idea what went on inside, or were so brainwashed by the Nazi propaganda. Or, they were desensitized by its knowledge that they casually went about their daily business. Human nature could be very revolting!

At the end of the lecture, I remarked. "There is a similarity between what I saw in the museum this afternoon and what just heard about a systematic neglect of the celebrated Theorem by mathematicians from Germany and other countries." Guinness described in his British 'muffled' accent how the Theorem published in 1931 in an Austrian journal of repute, was not mentioned in the works of distinguished logicians, Hahn (1933), Quine (1934), and MacLane (1934) to name a few! Curiously enough, Hahn was Gödel's PhD thesis supervisor in 1929 and had accepted his epochal paper for publication in 1931!

Other famous European mathematicians like Dieudonne, Hardy, and Bell did not include the Theorem in their books and monographs. The US mathematics was not on the international radar in 1930's! This reminds me of the *New York Times* in 1945 not publishing the Holocaust pictures and news reports coming out of the concentration camps! The Tsunami impact of the Theorem was eventually recognized like champagne bubbling out when the cork is removed. The Holocaust has transformed the surviving Jews into a nationalistic power in a new state of Israel born in 1948.

The argument that in 1930's only a few understood the Theorem, or it was of remote interest, does not hold as the top professionals in the field deliberately chose to ignore it. The lesser ones only followed suit. The intellectual traditions in the West (different from the Eastern traditions) build upon past references. Few top researchers have the time to understand others' works. Often the inaccuracies perpetuate in textbooks and research papers. **At times, research enterprise is no different from an intellectual Mafia.**

The Nazi Germany, or for that matter every authoritarian regime revises aspects of its history. They did not stop at the extermination of the Jews and the dissidents, but also undermined, minimized and distorted their intellectual achievements. There are hundreds of instances. The German Ministry of Education had the fullest control over the intellectuals, professional organizations, conferences, foreign visitors and invitees.

A lesson of history is; once a dictator wins over the intellectuals, then the masses are easily converted, or conversely. **There is only a small window of opportunity when individuals can speak up against an 'historic' wrong before it is too late.** If that opportunity is missed, then the voices of discord are snuffed and muzzled out ruthlessly. The price of freedom, like in the US today, is to be paid by constant vigilance, willingness to fight, and readiness to go at war.

The denial of recognition to Gödel's Theorem, or the holocaust of six million Jews is the ultimate price a society pays for not participating in its national politics. This is not the first holocaust in history. During the last one hundred years alone, the countries like Russia (under Stalin), China (Mao), Cambodia (Pol Pot) and India (British 1895-1905) have suffered from genocides far worse in magnitudes or brutality. Ninety minutes in the museum and ninety minutes in the lecture opened my eyes again to the dark past, present and future of mankind. **It tells what a single man can do, and what a man can undo!**

(April 13, 2006)

Skewness Of Linear Algebra

Last Thursday, there were only six students in my Linear Algebra (MATH 330) class that meets twice a week; six were absent! It was a small group of 14 on Day #1 that jumped to 16 during the Add/Drop week. Subsequently, three dropped out of the course/university and one has vanished away. The section, scheduled at 4 PM, draws mostly working students. The age varies from 21-40, and more than 50 % are juniors and seniors.

The empty seats sure dampen my enthusiasm for a moment. Being seasoned and believing in mixing 'fun' with mathematics, I try to play some David Letterman and Jay Leno. **After all, UNLV is situated in the Entertainment Capital of the World**! No entertainer in Las Vegas showrooms likes to perform to empty seats. Minutes before the show time, the audience is moved up to the empty pricier front seats. The back seats are darkened so that the entertainer remains under the illusion of a full house! In math courses, the back row seats are the preferred ones. That tells a lot about the two worlds!

Often high school students register in it as the schools are over by 2 PM. UNLV and CCSD have an Early Studies Program for the college bound kids. I miss them this semester, as during my previous two offerings (1999 & 2000), the top students were high schoolers. One is in a medical college and the other ready to go for math PhD.

A course, like Linear Algebra, did not exist when I attended college (1950's) in India. We covered a lot on theory of determinants (including its Laplace expansion) and worked on exotic problems from a classic textbook (by Hans Raj and Hukam Chand). Matrices were not a part of any course even at the master's level, though one instructor applied them in solid (three dimensional) geometry. Historically, the determinants preceded matrices.

The vector space approach to determinants and matrices is typically American as the British did not have it. The prerequisite is two semesters of calculus. The material begins to weigh heavier on students for its disconnectedness with the applied problems. They 'miss' the real world problems seen in precalculus and calculus courses. The linear algebra textbook has no such problems! Typically, it has dry drill and 'show/prove/find this' problems. **Telling the students that it has applications in sciences is not enough.**

Major wise, the students came from Electrical Engineering (6), Mathematics (4), Secondary Education (3) Computer Science (1), Physics (1), Economics (1). There is an equivalent course (MATH 365) emphasizing high speed matrix computation. 88 % have jobs. Two carried a full 12-credit load in addition to a fulltime job, and perhaps a family!

Dropping a course is no longer stigmatic. It reflects societal changes in values and norms. Students repeat courses for higher grades as well as for better grasp of the material. The growing number of non-traditional students in metropolitan schools like UNLV are re-defining withdrawals, retention and graduation rates. It is a different world of academe!

(April 17, 2006)

MADONNA AND MATHEMATICS ONLINE

"What are you watching?" I asked a colleague yesterday, while passing by her office, in the hallway. She was updating her online course material taped on a video. The video can be uploaded and downloaded on personal computers. Distance education, using high-tech, provides another delivery system in the fast pace of life today.

I have yet to teach a course in any format of distance education. Online education is a growing billion dollar industry. Major US universities are offering courses world wide, and people are attracted for their brand names. I used to think that some math courses can never be offered online. But three years ago, I was surprised to discover a South African university offering PhD in mathematics online! One does not have to go to South Africa.

I got a quick training in online instruction, but I did not try it. It was almost ten years ago when the state legislature earmarked two million dollars for incorporating internet in instruction. My reasons were weird. As I listened to my taped voice in the compressed video format, I did not 'like' it! Yesterday, I did compliment my colleague for her voice quality. Also, ten years ago, there was a technical glitch that would shift the picture from one site to the other by any audible sound even of a student's papers. I decided to wait for better technology, in future!

However, I continue to remain curious about online/distance instruction. This colleague responds to nearly 1000 e-mails from the students. Assuming five minutes to read and reply one e-mail, it means on the average an hour a day during a regular semester! The advantage is that an instructor can teach a course while vacationing any where as long as a PC is handy. Another instructor taught two online courses in a three-week session partly gone overseas! There are all kinds of juicy stories about the online instruction as are about the traditional lecture format.

At UNLV, only math courses below calculus are offered online. Whereas, in a regular instruction, one can only effectively handle 35 students, but online class size easily goes up to 70. Online courses make a sound business. There is a strong demand for them. And in America that drives the engine of higher education in public domain.

After the chat, my thoughts took a flight. I asked myself, "Why people are gladly paying up to $10,000 to watch Madonna live in her current *Confession Concerts* when they can buy her concert DVDs for ten bucks? A live concert, or play brings out a lot of emotions that turns into an orgasmic pleasure. A DVD can never do it even at its first viewing.

Mathematics is a paradigm, and I see my world through it. During a lecture, a seasoned instructor can weave every life yarn into mathematics, and conversely can spin mathematics into every aspect of life. Online clients are certainly deprived of this experience where the instruction is too clinical. But it suits people in rural communities and mega cities.

(May 27, 2006)

NEW CHALLENGE OF MATHEMATICS STUDENTS

"How many feet are in a mile?" shot out a student from the last row of a small auditorium type classroom. "Define your own mile and your solution will be accordingly graded," I quickly responded. No one else said a thing afterwards. I encourage solutions based upon different interpretations of a problem, particularly during a test.

Also, I take a little time to impress upon the importance of **definitions in mathematics**. Relatively speaking, definitions in mathematics are known for their consistency and least ambiguity, though to a naive some may be confusing and even contradictory. For instance, $1! = 1$ is OK to most undergraduates, but $0! = 1$ perplexes them as it 'naturally' implies that $1 = 0$. Another definition comes from exponents, $x^0 = 1$, when $x \neq 0$. They again wonder at both 1^1 and 1^0 being equal to 1.

The mile scenario above cropped up last Friday during a test in *Fundamentals of College Mathematics* course (MATH 120) that satisfies the math requirement in the General Education Core. The problem was stated as: *The dimensions of a soccer field are 345' by 223'. How many times a person must walk around the field in order to jog one mile?*

Here is the raw data. Only four out of 20 recalled the right number. The rest had 5281, 5260, 5360, 1582, 5680, 48,000, 3500, 60, 50 (by 3), 10000, 3645, 1290, 1000, 20. One simply put an answer without defining his mile! The late night comedian, Jay Leno, who often brings out trivia on public awareness of common issues, would love this story.

It is not right to link the wrong guesses with the prerequisites of this course. A mile is a part of US social lexicon. One could argue about it in countries where mile has been replaced with kilometer. In the US today, if some thing is not on the mobile phone or palm pilot, it is not in the mind either! In the attention shrinking and impatient modern societies, organic mind equals to the capacity of the plastic communication gadgets!

The students in this class are very non-traditional. I enjoy talking about life during a 10-minute break in 160-minute class everyday. More than 50% have fulltime jobs including responsible ones. Age wise, they are any where between two 18-year olds gradating from high school next month to two at least 40 years. More than 75% are females.

The heartening aspect is that the students are serious about learning. This may be their first and the last college math course. In fact, 50% are **seniors** who have postponed it to the very end! Their majors include dance, arts, nursing, psychology, special education, political science, counseling, communication, journalism, criminal Justice and social work. All of them started it with one common characteristic; math anxiety. I told them on Day One, that I may not be able to eliminate it in three weeks of summer session, but shall make it as an unforgettable experience of their life. I owe this challenge to myself, my students, my profession, and UNLV, the home of a unique blend of students in the world!

(May 29, 2006)

PRE-REQUISITES OVERSOLD

The issue of prerequisites in math courses has been a buzz and a bug for the last few years. Some believe that to remedy the dropout and retention rates, the prerequisites are crucial. The catalogs must have precise information. At UNLV, the Math Dept requires instructors to check the prerequisites on the first day. Failing to meet the prerequisites, the students may be administratively dropped. So far so good, if the policies are enforced.

But what about the students who have the prerequisites, but have forgotten the material as the courses were taken 'long' ago? Here is a scenario from my **Calculus I** (MATH 181) being taught in a 5-week session. The class meets four days a week from 5-7: 30 PM. On the second day, I set the following problems in a pop quiz:

Problem 1. For f(x) = ln x and g(x) = x^2 - 9, find the following:
(i) Domain and Range of f, (ii) Domain and Range of g, (iii) fog (f composed with g) and Domain of fog. Out of 20 students, only **one** got all the parts right!

Problem 2. Find the **inverse** of the function, $f(x) = \dfrac{x+1}{2x+1}$, assuming it exits. Only one (the same student) out of 20 had an idea, but could not find the inverse. Most thought inverse is another name of the **reciprocal** function. Incidentally, this student is local, but a finance major at USC where regular calculus is required. Ten years ago, UNLV's Bus. College removed a diluted calculus requirement, *Calculus for Business* (MATH 176).

This quiz material is covered in Precalculus I (MATH 126), whereas the prerequisite for Calculus is Precalculus II (MATH 127)! Do I have a dilemma? No; a problem, yes; of tailoring the material since every student satisfies the prerequisites. It is unethical and unprofessional to run these students out of the course by discouraging them. Nevertheless, such a quiz is essential to find out where the students stand.

A fact is, that math evaporates if math courses are not taken one after the other. That is the way I learnt math, but rarely does a student majors in math. Most are from biology, engineering, education including one in history. Everyone has a job and more than 50 % full time; some with families. These are the new realities of non-traditional students. I and most of my colleagues from foreign countries were full time students living with parents. It was an entirely different culture of learning. Here is a doctoral student in Chemistry who had two semesters of Calculus, 12 years ago. He has forgotten it, like I have forgotten much of my graduate math never used in research and teaching.

Calculus is all about Limits and its related concepts. Primarily, one needs mathematical maturity. The syllabi in all lower division courses are so tight that there is no room for reviewing, and it makes little difference any way! However, precalculus material is reviewed as needed in calculus. That is where, the experience, skills, and creativity are called upon on the part of the instructors. It is hazardous to preset a schedule of daily coverage. Awareness of new class realities can make a difference in students' learning.

(June 09, 2006)

IT IS ALL IN THE LIMIT

The Limit, in Calculus, is the profoundest and the most functional concept in the entire gamut of mathematics. Once Limit Chapter is covered, the students write a 300-word essay on it. A caveat is that it has to be written in the form of a letter to a close person. Letter is an ideal literary genre in which a person truthfully pours the heart out on a topic.

Yesterday, a pretty young girl, sitting in the second row with a skimpy summer skirt, did her reflex routine of opening up her legs and then crossing them over. For a matured mind, it was an enlightenment on Limit. An aha gushed out! The roving eyes are like 'x approaches a', and function values are the swelling slopes of thighs merging to the Limit F; as close to ε (epsilon) as you want! A college day's wisecrack, flashing from the labyrinths of time, said; call it, the Limit, Creation Hole, or the Whole Creation!

When mind is high on such a flight, then it finds convergence to the same limit irrespective of directions. The Muslims women in burquas may reveal only their toes, and that becomes a portal of entry into a woman's body for man's raving eyes.

However, the pendulous buns defy a limit, as in a 'neighborhood' the process is oscillating. The girls wearing skimpy tops and low pants provide the voyeuristic views for the limiting process. That is a professional advantage of working in colleges in the western culture. Such landscapes are taboos in many other cultures. The waist line goes Deep South that the elevated pubic region is the geodesics to the Limit. Sitting women flashing their sexy tattoos and rear cleavage provide one-way feast of the Limit. The flashes of pink undies extenuate the aroma and make the Limit delicious and heavenly!

One exception is when the eyes are frozen at the front cleavage and the heaving boobs make anyone ogling them breathless. The pertinent question is where is the Limit now? This is a perfect example when the left and the right limits exist, but the two are not equal! In fact, Heisenberg Principle of Indeterminacy creeps into the picture. You cannot see the two limits at the same time!

The bust dilemma is caused by the fact that the neighboring points are 'too far from a'. The function values have to be at points closer to the Point. One half of mathematics world is sitting on Limit. It is no wonder that the Limit manifests in most exotic settings. But one can only see them with a contemplative mind. The orgasmic experience after the intercourse is the ultimate realization, the Limit, of individual sensory limits.

Limit, like sex, has been heuristically appreciated by humans in every civilization. Just like the Hindu treatise of **Kamasutra** elevated sex into a fine art, a vehicle of fulfillment and self-realization (**Samadhi**), Cauchy, 100 years after Newton, formulated the ε-δ definition of mathematical limit. **Kamasutra** has an iconic stature in sensual delights. In mathematical delights, Cauchy's definition of Limit occupies the same status.

(June 16, 2006)

KNOW THY STUDENTS

"Are you also married?" casually I inquired from Pat who had 'survived' three out of five weeks of fast paced summer session. While holding a full time job, he took two solid courses including 4-credit *Calculus I* (MATH 181). "No, I had a fiancé once, who left as she could not keep pace with me." This small talk took place after the class was over. Another student, Bill, standing close chimed," I have a wife who can't keep pace with my life." He too has a fulltime job. Pat is nearly 30, and Bill 40.

More than 50% of the students have full time jobs. Of course, everyone has a part time job. **Working through college is an old American tradition**. A contrast, in life of a student in Las Vegas to any where else in the US, is at par with life in the US to one, say in India. This observation made 20 years ago is reaffirmed every semester.

Las Vegas is a different world of the students, and for the students. The other day, I drove my daughter-in-law to UNLV library. She was visiting us from Davis, a small university town in California. As she stepped out of the car, she was in disbelief on seeing a $60,000 convertible Mercedes with a student parking sticker. I have known the flashy life style of UNLV students for over 30 years.

After all, where in the world, a student can make $70,000 to $100,000 in one year? For the young, having good personality, attractive looks, athletic, or physique of body builders, well paying jobs in world-class gourmet restaurants, casino car parking, hundreds of classy or seedy night clubs are waiting there in plenty.

A natural question is: why do they seek higher education? It is generally out of lure for college diploma and American faith in life long learning. Over the years, I have closely known my students working as cocktail waitresses for high rollers and casino VIPs, topless dancers; male strippers, and legendary showgirls. In big casinos, waiters and parking valets easily earn $1000 in tips alone on a given night. They are also motivated students who want to be ready for another career when 'replacement' time comes in 8-10 years. **Those who live by the looks, die by the looks**.

An interesting phenomenon is that most students continue with casino jobs even after college degrees. **The average time for finishing bachelors is 7 years**. It shocked me in 1974 when I joined UNLV. Now I understand it. Traditionally, diplomas come first and jobs later; in Las Vegas, jobs first and diplomas later. My brother with a master's continues to work as waiter for over 25 years. A PhD, on the average, gets only $50,000!

It is essential to know the students before teaching them. It is important to know them before curricular programs are designed. It is pertinent to know them before the resources are allocated on recruitment, retention and advertising. With multitude of distractions in Las Vegas, the UNLV students are unique in the world.

(June 27, 2006)

ON A WITHERING DELIGHT

The graphing of a function is the pinnacle of differential calculus. **Like a musical symphony, it brings all the powers of differentiation together:** local/global extrema, increasing/decreasing, concavity/convexity, critical and inflection points of a function. Some precalculus concepts like asymptotes, intercepts, domain and range of a function get refined with the power of **Limit**. Sketching the graph of a function is like shopping in a Wal-Mart Super Center, where under one roof, one can buy any thing and service!

Yesterday was the middle of the third week, and also the middle of the 5-week summer session. After analyzing other aspects of the function, $f(x) = \ln(x^4 + 27)$, I said, "Let us now sketch its graph." Hardly had I finished saying it, when a few students were looking at its graph on their graphing calculators as they are getting 'smarter' and inexpensive.

I never prohibited any use of calculators. By choosing the right problems and re-phrasing the questions, it was possible to out-smart the calculators. Students thus saw how the power of calculators supplemented the understanding of mathematical concepts. There was now a feeling of inadequacy on sketching the graphs of 'normal' functions. While I was going through the motion of 10-points, the students followed them through their calculators. My patience too has become thin with graphing inside and outside the class.

These thoughts propelled me to my college years (1955-59) in India. Forget calculators, even the paper was expensive for doing math problems. However, in a seamless academic year (no semesters or quarters), these ten graphing points had all the time to get embedded in our minds. **The joy of understanding, as the graph revealed its shape in bits and pieces, was like watching a striptease dancer.** The limiting behavior of $f(x)$ *and* $f'(x)$ gave thrilling moments. Often, the final product was beyond any initial guess on the graph.

'You have come a long way, Baby!' For the last couple of years, the time spent on curve sketching has been diminishing. Now more time is spent on the evergreen related rate and optimization problems. Also, more students are coming from biology and biochemistry. Calculus is no longer monopolized by math and engineering majors. During the last four offerings, my section (out of 8 sections) of **Calculus I** (MATH 181) did not have even a single math major. This session has only 3 from engineering. The remaining 17 are from diverse disciplines including two from nursing and one from history!

Student diversity is a sign of mathematics progress. Concepts are also prematurely born and buried. In atomic structure (chemistry), I was taught (during 1950's) the notion of valency; that no longer exists in textbooks! The modes of transportation, designs of homes, and even life values change in a span of 50 years. Yet, there are some fixed points of a function, and of life in general too. The calculus concepts of limit, continuity, differentiation and integration will live forever!

(June 22, 2006)

I Was A 'Professional' Student

Comparing my college life during 1950's of India with that of UNLV students, I was a professional student! Full time description does not capture the daily routines, we had. Imagine the life of a professional baseball player playing 162 games or a basketball player playing 82 games with two months of training and playoffs. etc. There is no life beyond the game for solid 10 months in a year. Whereas, the US professional players are reasonably rich, we, the students in India, studied to escape out of poverty. We lived with our parents who supported financially all our basic needs and college fees.

The academic year was not split into semesters or quarters. The comprehensive exams were held at the end of two years covering the material since Day # 1. Reviewing problems, 4-5 times before the Final, was normal. A rival classmate, Subhash literally studied 15-16 hours a day for nine months. He once told me, how he considered his time wasted when he went up to greet his father returning home from an out-of-town trip.

There was no question of sports and play, as waking hours meant studying. The boys, in particular, pampered over girls, never helped in any household chores. Being a student added to the privileges! Memorization was a key to the success; there is no short cut to time it takes. Another friend memorized hundreds of pages of books and crafted notes, so that he doesn't have to think during the exams, each lasting for three hours.

In the prevalent US system, instructors are apologetic when they expect their students to memorize a simple formula, like quadratic formula. That is why the trig identities and calculus derivatives create panic amongst the students. Generally, a 3x5 or 5x7 card is permitted, or a list of formulas included in a test. Some instructors give out a practice test that is 90% similar to the Final. Since there is no time to review or go in depth, such practices are now understandable.

To a large extent, the US college athletes are nearly professional athletes, the way they spend time in their sports on and off the field. The big difference is that back in India, we were not paid any thing or discounts for studying. The great hope was that education was an exit gate out of poverty Indian society was steeped in.

Let me add, that in towns and cities, there were few full/part time employment opportunities. The paradox of that life of the have-nots was that working for money, while as student, was considered a social stigma! In college, I refused to tutor students even for an hour. The idea being, spend that one hour in reviewing and memorizing.

It is amazing what my life has witnessed; from four years of college in Bathinda situated in the most backward region of Punjab to watching students in the glitzy town, Las Vegas. In one place, there was hardly any distraction to studies, and in the other, it is all about sensory distractions of any scales. It has tremendously enriched my life!

(July 2006)

ON INTERACTION WITH STUDENTS

"How are your students today?" asked a colleague in the hallways. It was the last day of the 5-week summer session. "My students feel the way I feel, shouldn't they?" I responded after a little pause. Usually, I like to throw a twist in answering a generic question, as it makes the other think for a moment. Well, that response was like a curve ball.

Elaborating it, I said, "In any sports, the players are coached in the mould of their coaches. A well coached player thinks and plays like his/her coach. There is a process of re-making the personality. I wish we could bring that physical and mental training into our classrooms." He smiled, and we walked to our respective classes that meet for 150 minutes without break, Monday through Thursday afternoons.

However, it is too much to expect mental transformation in five weeks. For the football players, when the regular games are played during the fall semester, the training begins in spring and goes for 2-3 months. The players are selected to fit the profiles of prominent coaches. Sport psychologists, in athletic departments, screen players not just on their athletic abilities and skills, but also on their willingness to follow the philosophy and regimen prescribed by the Head Coach. It is true whether the game is played in high schools, colleges or professionally. It is one seamless process.

At UNLV, undergraduate students try to fit their courses around their jobs. Taking a course from a particular instructor is only incidental. Also, 30 meetings in 15 weeks is too short a period for any rapport to develop between the students and instructors. Most instructors don't care to know the names of their students, if the instruction is non-interactive. Student often recall their instructors, not by names, but by their ethnicity, like Chinese, Asian etc.

During four years of my college in India, I had only two math instructors. Both have made tremendous impact on my life, though they were diametrically opposite!. Each one brought out a different strength out of me. In the US, such an interaction between a student and faculty member takes place at the graduate level when a student chooses a faculty member as his/her thesis advisor after taking at least 3-4 courses.

In the US, the philosophy of the undergraduate education is more of generalists rather than of specialists, as far as the curriculum is concerned. The wide range of course work can be compared to a buffet where hundreds of food items are served. The strategy is to sample as many items as possible, and go for the seconds that are liked. In contrast, during the last two years (1957-59) of my college in India, 60 % of course work was in math and 30 % in English! A student with such a bachelor's is solidly prepared for PhD in any university.

(June 07, 2006)

WAR SPARES NONE

Umakanth, 5', 3" weighing hardly 90 Lbs, is a shy 25-year old graduate student. If he were suddenly teleported in a battle front of hot Iraq, then he would literally melt away within an hour. But sitting in the air-conditioned facilities at UNLV, he is making a small contribution for the US to win this war. I just attended his successful MS thesis defense.

Professors Brendon and Mohammed, his co-thesis advisors, spoke high of his research. They started researching on armed vehicles in the Dept of Mechanical Engineering, and subsequently received funds from the Army Research Laboratories. All the branches of the US defense forces promote basic as well as applied research in various universities across the country. It widens the base of professional research and public interest.

The focus of this engineering problem is the skeleton design of armored vehicles using aluminum. We all read how in Iraq, army vehicles are attacked by rockets, and detonated by blasts hidden in the ground. First, the problem is mathematized as a boundary value problem of wave propagation. The applied mathematicians and statisticians develop and refine mathematical models that are further used by scientists and engineers.

A key to victory in war is minimum time taken in bringing improved weapons and support accessories from laboratories to factories, and to the battlefields. There are hundreds of foreign graduate students from countries like India, who provide an intellectual labor force in the development of new weapons. Umakanth's simulated research, being a part of wider research, may be pushed for lab testing. Once the engineering design is finalized, it goes for mass production. On the assembly line, the blue collar workers will have no idea how and where, the new design became a reality.

Another thought was, that really no one stays unaffected in war. During war, the economy has the same highs and lows as in peace time. Intellectually, each brings out different aspects of human excellence. The interesting thing is, like in politics, the science and engineering problems cut across nations.

From his references, I noted a few MS theses and PhD dissertations on related topics done at the universities in Sweden and Canada. Knowledge never had boundaries, though men and women in sciences do have national allegiance. By classifying research, a nation maintains its edge over the enemy. Since time immemorial, intellectual espionage has gone on even during peace time. No wonders, peace is called a preparation for war!

Often I ask the Indian students, how coming to the US has intellectually affected them. Invariably, the answer is the research experiences that even the best Indian institutions, like IITs don't offer. One day, another Umakanth may help US Army design a vehicle, that when bounced off the road after explosion, goes into a rickety run, and lands back on the ground without significant damage to its electronic system, driver, and commander!

(Aug 07, 2006)

ON MATHEMATICAL REFLECTIONS

Three years ago, I started writing tidbits based upon my professional life spent in teaching mathematics in the colleges and universities in India, USA and Malaysia. The more I wrote, the deeper and clearer became my ideas, and better came out the new pieces. A year later, I started calling them *Mathematical Reflections*.

A couple of days ago, I received an e-mail from a journal editor that my 'master piece' *Reflection, Calculus Defines Civilization* will appear in the **Class Notes** section of the journal. The Mathematical Association of America has published a couple of them in the opinion space of the *FOCUS*.

While on a recent walking beat, a question surfaced up, as to what do I myself think of these *Reflections*. Class notes mean the ideas coming out of the pedagogical observations and fit for class environment. My *Reflections* are sparked by remarks of students in the classes, conversations with colleagues in the hallways and offices, and attending meetings and conferences. General readings also provide new ideas.

Yes, my students love these *Reflections* so much, that I distribute a bunch of them in my courses. They want them published as an accessory to their textbooks. But having no experience in publication, this project remains on a back burner. They will have a published life too! Currently, there are at nearly 100 math professionals, worldwide, on my e-mailing list who receive them, appreciate them, and look forward to reading them.

My forte in writing these *Reflections* is my uncanny ability to overarch two distant pillars in life. There is a dash of history, as history of ideas and ideologies has been my passion. Curriculum history of the US and India are braided with personal touch. The curricular changes in the US are continuous. In developing countries, like India, they have jump discontinuities and vary from state to state. The US is McDonalds and Wal-Mart country.

The language of **Reflections** is not terse, dry, or filled with technical jargon. Even during lectures, I try using flowery expressions, humorous anecdotes and unconventional phrases connecting mathematics with life around. The subliminal themes of my courses are: **Think Mathematically** and **Life is Mathematical.** The problems of life including religion and politics are resolvable, once they are mathematized! That is where I stand.

There is a variety amongst the **Reflections**. Some are actually X-rated, meant for those deep into applied mathematics, and can relate mathematical concepts with the ultimate sensory pleasures. A few are literally nuggets, some thought provoking, but each one engages the mind. Even a mundane topic gets a spin, touch of salsa, and spice. **If one has consciously gone very deep into any endeavor of life, or touched its highest pinnacles, then that experience is relatable with any other aspect of life!**

(Aug 17, 2006)

A CONVERGENCE OF
SLAVERY AND MATHEMATICS

"What shall I gain from learning mathematics?" is a perennial question the students often ask out of curiosity, ignorance, or confusion. When this question was posed to **Euclid** teaching in the famous Greek Academy founded by Pythagoras, he instructed his slave, *"Give him a coin if he must profit from what he learns."* Yesterday, after reading this popular historical anecdote in Stillwell's book, *Mathematics and its History*, I stopped to ponder it over.

Great empires are also built on the brawns of the slaves. After consolidating the empire Alexander inherited, he expanded it all the way to the east in the western fringes of India. Depending upon the social conditions, the local population was accordingly subjugated and ruled. Two centuries later, the Romans, in their heydays, continued slavery as the movie *Gladiators* captures it from the sporting angle. The Arabs enslaved the Africans and infidels long before the Europeans shipped slaves over Atlantic in the 16th century.

Thomas Jefferson's contributions to the American public life and politics are very well documented. His intellectual prowess was phenomenal. However, he maintained a stable of slaves. A recent movie has brought out his relationship with a black slave girl and fathering kids with her. The historical fact is that slavery lies in the foundations of America. **A new empire is built, and it flourishes only, when the society conquers. The dilemma is that the conquest over men is generally a part of the total conquest of ideas in science, art, and literature.** The great American ideas and ideals of the 18th and 19th centuries are nourished from the great ideas of ancient Greeks and Romans!

Well, my second thought from Euclid's quote was that in every age there are individuals who pursue knowledge for the sake of knowledge. The story of Grigory Perelman of Russia, also hitting the internet yesterday, is of a modern Euclid. His declining the Fields Medal and $15,000 must make Euclid smile in his grave!

No matter how abstract a mathematical, philosophical, or literary work is, some one is directly or indirectly supporting the genius for his daily bread and butter. Some geniuses may not be good at working in teams, or feel comfortable in lime light. However, a stern lesson of history is that in the moments of survival of the society, as a whole, an individual must toe the collective line until the danger is over.

The controversy of pure and applied knowledge (mathematics) can be traced to the eons of time. Aristotle was highly applied. He tutored Alexander the Great, and along with his nephew Calisthenes counseled him during his conquests in far East. **The WWII was first won on the blackboards and in the laboratories of USA!**

(Aug 24, 2006)

MY AFFAIR WITH DIFFERENTIAL EQUATIONS

"Welcome to my world of Differential Equations (DE)!" That is how I greeted my students on the first day of this course (MATH 427). DE stir memories going back to 1959, when I started my master's. On examining the US curriculum, my bachelor's and master's from Panjab University, Chandigarh (India) were indeed in mathematical physics; 50% pure theoretical physics and 50 % mathematics; pure and applied.

DE formed 25% of a one-year course (called paper) that included Elementary Real Analysis (RA) for the remaining 75 %. A 3-hour comprehensive exam at the end of the year determined it all. All the class tests, home works etc. carried zip! The final exams, written by external examiners, proctored and graded by others from different institutions, were very confidential. It also reflected how well the instructors prepared their students.

It was an era when buying textbooks was beyond the means of most students. We relied on good lecture notes. I seldom kept up during the class lectures, as would catch up later. After a month or two, I started losing interest in RA. On the top, a pretty classmate, Swaran Kashyap sitting in the front row understanding all, really fueled my frustration!

It is interesting to recall the instructors. Real analysis was taught by I S Luther, a fresh PhD in number theory from Illinois. DE was charged to D. D. Joshi, another fresh PhD in statistics from Paris. During the very first classes, Joshi confessed that he was out of touch with DE. Half of the class never came. He spent more time looking at a problem than working on it. **You never stare a person or even a math problem for too long!**

Joshi was a thorough gentleman, and out of respect, I never missed his class. It paid off huge one day, when he actually solved a DE that I could not do it for a few days. That moment is vivid, as if I learnt how to think! (In 1966, he wrote a reference letter for me to Indiana

University) When Dec class test on RA blew me away, I started to strategize for the final exam. A classmate, Subhash was in the same predicament. Also, known to each other since the 5th Grade, we decided to study DE together.

The final exam problems were divided into two sections. The one, on DE, contained three problems and the other, on RA, nine. The only condition was that for a total of at most six problems, at least one was to be attempted from each section. Our goal was to knock out all three problems on DE. We searched library books, and hunted for DE. During the 1950's, the theory of DE was not that much emphasized. With subsequent mastery over the techniques, I coined a phrase that **'I can smell a solution of a differential equation'**!

Yes, I took care of all the three DE problems, and did not-so-bad in the RA portion to score 80 %. In that system, any score above 33% was considered high! In RA, I worked on a maxim; **do little, but do it very well,** when the odds against are high. DE was my savior, and the love for it, eventually became my bread and butter! That is another story.

(Sep 18, 2006)

When Telling Means Knowing

"You are invited to give a 5-minute speech to a group of bright high school math students. Your topic is the *First Order Differential Equations*. List the main points that will make a positive impression." This was the last problem in a test that I gave this week, in the first course on Differential Equations I (MATH 427). Though this course has been taught before, but such a question was never included. However, a number of reasons favor it now!

Sharing a good experience is irresistible in every walk of life; it is human. People love talking about **gastronomical** delights. We inform friends about the **visual** experiences of going to a museum like Smithsonian. The ladies love to share (**olfactory**) new perfumes. People proudly pronounce the joy of (**auditory**) musical concerts and symphonies. Whether communication is verbal or written, the organization of the matter comes out of its understanding, and conversely.

The ability to communicate a technical subject is a hallmark of its mastery. In democratic societies, the experts often face non-experts like legislatures and community leaders on technical public issues. The emphasis in technical writing started 25 years ago and students in science, engineering and business were required to take a course in writing. The Honors College at UNLV lays special emphasis on writing in the entire program. While teaching Honors math courses and seminars, my students have written several short papers. A letter to Dear One is the most successful format as in a silent monologue one can pour one's heart out. For example, the calculus students wrote letters on **Limit, Differentiation, and Integration.**

There is a renewed emphasis on writing in all the courses. Last year, I attended a half day writing workshop sponsored by the English Department, Teaching and Learning Center and General Education Committee. The basic idea is that writing should be an integral part of every course. It reminds me of a famous quote (???) it said that **writing makes a person perfect** (in the grasping of an idea).

In any math course, there are several topics where students and instructors talk about their difficulties, uselessness or usefulness. They all provide perfect scenarios for short writing assignments. Above all, it is a breath of fresh air from mundane homework assignments and routine test problems.

Grading this problem was fun. A few struggled to write even 4-5 lines, but a couple of students filled a whole page. Yes, it was announced that such a problem shall be on the test, and nearly 25 pages of the textbook were suggested for reading. Well, the result is promising enough to justify its inclusion in the second test too.

(Sep 27, 2006)

A PERSPECTIVE ON RESEARCH

When it comes to problem solving, 'Two heads are better than one, is a cliché in the US public life. It applies not just for figuring out mundane problems, but in fundamental researches too. Mathematics, in general, is no exception, though research in 'pure mathematics' is still regarded a highly individualistic intellectual activity.

A thesis is that the freedom to think and challenge the present is a function of deeper values imbibed from one's religion. The Hindu religion places paramount importance to individuality. Consequently, while growing up in India, a belief set in my mind that basic research is also a lone man's pursuit! It is indeed more of a team work.

The US life has opened my eyes. Intellectual collaboration between two individuals is like free for all popular fights using legs, fists, and even certain objects. This is witnessed amongst the US researchers. They argue over a problem, but at the end of the week drink over it. This cord struck my mind yesterday while reading a brief article on the Bernoullis that dominated the world of mathematics spanning three generations (1650-1800). Out of a total of eight Bernoulli mathematicians, three all time greats are; two brothers Jakobi (1654-1705) and Johann (1667-1754), and Johann's son, Daniel (1700-1782). Jakobi was the first child and Johann the 10th born.

The lives of the Bernoullis are filled with stories of competitiveness and jealousy between brothers, cousins, uncles and nephews, and ultimately between father and son. During the beginning years, the youngsters did learn mathematics from their elders, but as professionals they became fiercely independent and arrogant. It is like the Mughals ruling North India for nearly 200 years (1525-1725). The heir apparent brothers were killing and fighting all the time for the crown of Delhi. The Chinese revolutionary, Mao Tse-tung (1893-1976) said, "The power comes from the barrel of the gun." It applies in any walk of life, if one aspires to be at the Everest of its mountain.

Twenty years ago, a white American physics colleague frankly confided how he regarded Indians as not-so-good research collaborators. His main reason was that the Indians keep formal relations with their colleagues and address them as Dr Josh etc.! A free exchange of ideas is not possible in official relationships. It is like doing gardening work wearing a 3-piece suit. There is indeed a mental framework for collaborative research.

I have been a victim of this conditioning of mind since my Hindu upbringing. Though now in my 60's, I can't help maintaining a prescribed distance with my erstwhile teachers and older relatives in India. They also can't shake off my 16-20 year old image known to them for the first time. Consequently, we deprive each other the exchange of accumulated wisdom. Forget the insipid phone conversations on rare meetings.

Actually, my story is one of its kinds. Despite frankness on the part of my PhD thesis advisor, Robert P. Gilbert, I have yet to address him as Bob! He is 8 years older than I. For similar reasons, my collaboration with any one on a scholarly project has not taken place. If I were to start academics all over again, teamwork will be high on my research agenda.

Amongst the Hindus, touching other's feet displays the highest regard. At the other extreme, it is a sign of ultimate physical, mental and spiritual subservience. No other culture demands it. No wonders, it has set the Hindus apart from each other. They are neither together in the battles of ideas, nor of weapons! The price they have paid is the highest in history; 1000-year of foreign subjugation, from the 11th century to 1947.

Amongst the Indians, the age gap is unbridgeable, as they show disproportionate respect to age and status. In the present American culture, a six year old is introduced to a 60 year old on a first name basis. It really offended me initially. But it removes any fear of superiority from the minds of the youngsters. The epochal ideas seldom come from obedient and law abiding persons in any walk of life.

During the corporate brain storming sessions, the CEO and low level executives, all are equal! There is a place for the formal address. Appropriate respect is due to an office, but a person has to earn the respect from others every day.

(From Daniel Bernoulli's 1743 letter to Euler) "Of my entire *Hydrodynamics*, not one iota of which do in fact I owe to my father. I am all at once robbed completely and lose this in one moment the fruits of the work of ten years. All propositions are talked from my *Hydrodynamics*, and then my father calls his writings *Hydraulics, now for the first time disclosed*, 1732, since my *Hydrodynamics* was printed only in 1738."

Daniel claimed that his father, Johann actually published his work in 1743 and predated it to 1732. There are two acts of commission; one of plagiarism and the other of distortion of facts. However, it is Daniel's side of the story. More research may be needed to find Johann's side. Incidentally, it was Johann who for a regular exchange of money, let a wealthy merchant (perhaps frustrated mathematician) Marquis Hopital publish his research work under Hopital's name (of Hopital's Rule in calculus). Was Johann not prostituting research? Free riders are very common on research papers these days.

Nevertheless, my focal point is not plagiarism or distortion of facts, but for 'fighting' in research for supremacy and a piece of immortality. Like they say; every thing is fair in love and war. War has infinite forms. The research that the Bernoullis generated, the new areas they opened up, and the impact they had on science and early industrialization are humongous. Europe dominated the world through the end of the 19th century.

My conclusions on research are aptly summed up by the remarks of Andres Liljas of the 2006 Nobel Committee for Chemistry on the Americans sweeping all the 2006 science Nobel Prizes this week:

"....Besides a supportive research granting system in the US.....
American universities often have more creative environment than
in other countries...Creative means that people interact with each
other a lot. It means you should talk with each other also, and not
work as hermits, separately."

(Oct 04, 2006)

TIME TO RE-EXAMINE

My mind started racing while grading a Differential Equation (MATH 427) quiz given yesterday. Out of three problems, no one got a single one right! It was based on the material covered during the previous meeting, two days earlier. In fact, the problems were similar to the ones discussed in the class. A handout had solutions of two of them!

Out of 24 students, half are math majors. All are juniors and seniors. My first reaction took me to the world of athletics. There are 'punishment' stories when a team performs very poorly. For missing a free throw, the player may be ordered to shoot 2000 times. The players run miles and lift weighs till they literally drop. If the coaching has been the reason for terrible performance, then coaches are routinely fired during the middle of a season.

This is not the first time that the students have performed poorly in my course. It does make me re-examine my presentation and do an analysis of the whole situation. Otherwise, it is embarrassing to cover the new material. The US students are not a captive audience like in India, where instructors can ram the material down their throats.

Two days earlier, on Tuesday, a pop quiz had surprised the students, as all the previous quizzes were given on Thursdays. The quizzes, taking around five minutes, keep the students on top for the ongoing material. Perhaps, no one expected the next pop quiz immediately on the following Thursday! "I do not like to be predictable!" I remarked. Besides a **Mathematical *MANTRA*,** a good problem a day keeps the math anxiety away, I strongly recommend students studying for twice the time on the day of the instruction.

The students are all working, and 25% full time. One is taking four 400-level math courses and running between two jobs! He has not missed a class and is not failing at this point. Back in India of 1950's, being a student meant, you only study (little playtime) during the waking hours, not helping in any domestic chores. Forget jobs; there were none! We lived off our parents. In Las Vegas, students have the opportunities to earn more than $50,000 in some casino jobs, I admire them for finishing BA/BS in 6-7 years.

However, my disappointment was short lived, as I read an article on the American sweep of 2006 Nobel Prizes in sciences. It observed that more Americans have been winning the Prizes, as the US has the largest base of science researchers. The US university environment is far more creative than any where in the world. Nothing to worry now!

In Las Vegas, there are stores, where one can buy a good shirt under ten dollars to another for a thousand dollar! Math majors are products too. From varied arrays of math students in over 2000 colleges and universities, some would easily knock out my exams. Seven years ago, I had this experience of a graduate from a prep academy who took all my Linear Algebra exams in one shot, and lost only one point! **The strength of the US education system lies in its accessibility, flexibility and continuous updating.**

(Oct 06, 2006)

GAUSS, MATH AND CHAIR

"**Teaching is hindrance in doing research.**" I heard this quote 30 years ago from a colleague who was then a rising math researcher. Since then I have tested this hypothesis numerous times, and statistically speaking, it is true in 98 % of the cases modulo what defines a researcher in a specific institution. This remark is attributed to Carl Friedrich Gauss (1777-1855) acclaimed as the greatest mathematician of the 19th century.

However, last Monday a topical investigation brought another side of Gauss in focus.   *"Gottingen in 1846 was not the Mecca for mathematicians as one would have expected it to be with the great Gauss in the chair of mathematics. Professors kept aloof from students and did not encourage original thinking or lecture on current research. Even Gauss himself taught only elementary courses. After a year, Riemann transferred to the University of Berlin, where the atmosphere was more democratic and where Jacobi, Dirichlet, Steiner, and Eisenstein shared their latest ideas."* (Stillwell's *Mathematics and Its History*; p 290) For his varied fundamental researches, Gauss was known as prince amongst mathematicians. Gottingen University one of the oldest universities in Europe that during 1930's of the Third Reich was completely purged of the Jewish faculty including Albert Einstein!

It stirred many personal memories of the leaders of math departments. Time is ripe to put them on record. The distinction between department chair and head is thin. Generally, Chair is a faculty member internally elected for a 2-3 year term versus Head appointed from outside for a longer term and given more 'authority'. **It is an American administrative innovation in higher education.** The purpose behind Head is to infuse new ideas and leadership in a dept stalemated, in tailspin, dysfunctional, and not serving the students and community. This administrative philosophy is one of the several factors that have made the US university environment as the best in the world.

Indian universities modeled after the British colonizers had 1-4-8 model of a typical math dept with one professor at the helm (4 associate/readers and 8 assistants/lecturers). Professor and Head usually changed only on retirement or death. In India, since 1980's, not only the headships/chairmanship rotated by seniority (not election), but also the number of (full) professorships go by qualifications rather than by any ceiling on a rank. **I call it an American seismic effect in the world of higher education.**

Of all the variables that go into the making of a good math department, chairmanship or headship (used interchangeably from now onwards) is the most crucial piece of the puzzle. The total academic environment of a department is directly proportional to the persona of its chair. The 'Gauss' moment has compelled me to roll back my mathematical video. Since 1959, as a graduate student and faculty member in India and USA, I have observed and dealt with a total of 11 department chairs. **Whereas, the chairs in the US govern with faculty consent, in Asian countries, by and large, it is by the authority from top to bottom.** It is a reflection of the political systems, as well as their traditional social structures.

The standout chair is S. D. Chopra (PhD, Mathematical Seismology, Cambridge, UK) who during his tenure (1962-78) built a reputed Math Dept of Kurukshetra University newly started in the state of Haryana. The state had no intellectual atmosphere or traditions, but has been famous for its martial history in providing the best soldiers to the country! Despite limited state resources and parochial campus, Chopra nursed and nurtured the Department in two other specialties far removed from his own. **That is one hallmark of an effective chair.**

George Springer (1967-70), an analyst, turned Indiana University (IU) Math Dept as Number One in functional analysis (operator theory). Besides hiring promising young faculty, he pulled away star operator theorist, P. R. Halmos (1916-2006) from the University of Hawaii. As a speaker, researcher, teacher, mentor, and author of textbooks, monographs, and articles on math education, Halmos was one of the greatest American mathematicians of his time. However, his teaching style and the way I grew up learning math were orthogonal. But at IU, there used to be a waiting list for his courses!

In this context, it is worth mentioning that Halmos also had a stint as a chair of the Math Dept in Hawaii. He noted it is his popular Automathography, *I want to be a Mathematician* how he was not cut out for it. Though he did complete the 9-month term as chair, but after three months into it, he had decided to leave Hawaii for IU! **Chairmanship and quality research do not go together.** It is pertinent to be clear; to be or not to be a chair.

The chairmanship has perks including financial, and opportunities for upward mobility in the administration. But if one does not have interpersonal skills of communication, understanding of personnel and disciplinic differences, then the tenure may be rough and tough. Graduate schools train only for research, provide some exposure in teaching, but do nothing about administration. It is a paradox; one learns it all into the position.

By the very individualistic nature of mathematics, one is publicly withdrawn and less outreaching. The odds are against a mathematician in cracking into top echelon of higher administration. However, with legislative demands and public scrutiny on institutional accountability, assessment, diversity, and various compliances, there are far more openings in administration than ever before. Twenty years ago at UNLV, there were no layers of administrators between the chair and the dean, or between the dean and the provost. Today, there are several attractive positions of assistant/associate deans/vice presidents/provosts.

Whether in the US school districts, or in the universities, the biggest growth in hi-fi jobs is in the administration. Unless, one has a vision to move beyond the department chairmanship, dept chairing may not be a very satisfying experience at the end of the day. **A good chair mentors and nurtures a few faculty members for administrative roles while himself making waves for upward mobility.**

(Oct 28, 2006)

P. R. Halmos, As I Remember
(March 03, 1916 - October 02, 2006)

A measure of a great life is its aperiodic seismic effects over a long time. Call it a sheer coincidence, or an act of premonition last week, that from the internet, I learnt about Halmos checking out from the planet Earth and moving elsewhere. Halmos was not a gypsy like his Hungarian mathematician friend Paul Erdos (1913-96), but he did have a craving for moving and visiting different universities (at least 18) and countries every few years. It is reflected in his intellectual vitality. Personally, since starting professional life in 1961, I have already moved ten times.

I first saw Halmos at a spring 1969 colloquium he gave before joining Indiana University (IU) in the fall. The colloquium room was hastily changed twice to accommodate a surging audience from all over the campus! Halmos lived up to the 'billing'. With uncharacteristic personality, combined with clear exposition of his research topic, he won over the math experts and **naives** alike.

Before seeing him in person, I had heard of his name from his internationally popular books, in India. As a matter of fact, I was advised to bring along the Asian editions of a few books including *Measure Theory and Finite Dimensional Vector Spaces*. The cheap Asian editions of science and math books helped Indians and Chinese, in particular, to raise and revise their colonial math curricula. Halmos was a scintillating writer. For clarity and sharpness, he flexed the language and developed a snappish style that is rightfully Halmosian. **Halmosian is a newly minted 'coin' commemorating his persona.**

In fall 1969, three sections of real analysis (Roydan based) were scheduled, and everyone wanted Halmos! I could not get into his section. He taught courses in a progressive cycle in order to groom prospective students for doctoral work. I finished real analysis with another instructor, but in fall 1970, I took Halmos' functional analysis course (there being only one section). His teaching also extended down to the freshmen level where he loved teaching a *Math Appreciation* course.

Halmos proclaimed his teaching approach after a famed Texan mathematician, R. L. Moore (1882-1974). But I think it was all **Halmosian**. He perhaps never met Moore and supposedly heard stories of his style. Twice, during a semester, Halmos would give out 4-5 sheets of problems including standard theorems and explain how to present them in the class. He would sit at the back of the class, and interject his comments and questions as students presented individual problems. Being conditioned to learn by lecturing for years in India, I was an '**invariant subspace of Halmosian** operator'!

Of all the math instructors in my life, I rank number theorist, H. R. Gupta (1902-1988) of Panjab University (PU) Chandigarh at the top. He used to involve all the students in solving a problem during an hour. It was an assembly line approach of finishing a complex product that no one person could complete it by oneself. However, Halmos pumped and filtered the students from the point of view of their research capabilities.

A common feature between Halmos and Gupta is that they not only knew their students by name, but also called upon them during class participation. I complimented Halmos for pronouncing my last name, Bhatnagar, perfectly, as the syllable '**bh**' is phonetically very difficult for the Americans. He impressed me when he crisply pronounced another Indian name, **Srinivasan** who taught complex analysis during 1960-61 when I was at PU. One day, in the context of a theorem, Halmos remarked how T. P. Srinivasan was known in the mathematics community for providing alternate elegant proofs of theorems recently published. Of course, in math, it is the first proof that counts for a piece of immortality. But beauty and elegance are also the attributes of a mathematical proof!

Halmos was dedicated to the teaching of his classes. As a famed mathematician, he constantly received invitations from all over the world. Being conscientious of his obligations towards his courses and research, **he accepted only one invitation per semester**! On the first day of classes, he used to inform of the day when he would be gone. However, in his absence, the students carried on with the class business.

It reminds me of an eminent professor from India currently on a 3-week overseas trip for attending a 3-day conference in Las Vegas! I was surprised when he casually mentioned it over lunch. Neither every US professor is like Halmos, nor is every Indian professor like this case. Again, I recall H R Gupta who once politely declined a luncheon invitation with India's first Prime Minister, Nehru on a short visit to the PU campus. The reason? The lunch hour clashed with his class hour! I have tried to live up to this benchmark.

Halmos and his wife Virginia were married for 61 years. To the best of my knowledge, they had no children. While he was always appropriately dressed, his wife wore longish gowns with disheveled hair, the style of the pop culture and Vietnam protest era. Later on, I learnt that she was a scholar of philosophy and Latin literature! Nevertheless, **they must have enriched each other's lives to stay married for six decades**. Such longevity in marriage is unthinkable whether in India, or the US of today.

Halmos and his wife used to come to Hoosier Courts, a married students housing complex where we lived on the IU campus. They carried two golden cages with a cat in each for cat sitting done by the wife a math graduate student. He gladly paid $1.00 an hour when charges for human baby sitting were 50 cents/hour! A story about Halmos' cats was that they were trained to play table tennis. **Creativity never stops and finds other outlets**. H R Gupta had developed a manual technique for making espresso coffee!

One common trait between Halmos and I is the love for walking with a stick in one hand. For many years in the US, I walked at least 3 miles back and forth to the university. Halmos was addicted to walking. He often parked his car 2-3 miles away from his office and walked. On certain days, when he was in a mood to walk 10-

12 miles, his wife used to drop him off at campus and pick him up at some other point. There were no cell phones during those days. Halmos belonged to a tradition of great Greek philosophers from the era of Socrates and Plato who mused, taught, and conversed while walking. Most of my *Reflections* also emerge during my walks. I love 2-hour walks, but never the brisk types that Halmos took.

Just like his angular athletic face, his personality was angular too. Halmos never hesitated to rub shoulders with people around him. In every group, either people loved and adored him, or just hated and felt like punching him. Here the **set** of the middles was **empty**! During my IU days, his argumentative bouts with a couple of colleagues were known to everyone. Consequently, they never greeted or acknowledged each other in the hallways or going up and down the staircase. **A personality is equally defined by the type of enemies one makes.**

The sharp differences of opinions also showed up in his 'opposition' to faculty engaged in numerical analysis, applied mathematics and statistics. The title of his books and articles bring out his provocating and stimulating personality. A couple of his catchy titles are: *Applied Mathematics is Bad Mathematics, Thrills of Abstraction* and his *Automathography: I Want to be a Mathematician.* I remember him driving a red sports car, and often wearing sport jackets, or t-shirts.

Halmos was the best PR (Public Relations) person of mathematics of his time. He was confident, witty and cocky with interviewers. Once he told the class that no matter what public mathematics lecture you give, **prove at least one theorem.** It is tough to meet this standard, but is not forgotten.

For years, Halmos was a **fixed point** during the January Joint Meetings and seen surrounded by people chatting on some mathematics problems. During a class, he stressed upon joining the AMS or MAA. There were no other math organizations in the 1960's. It prompted me to join the MAA in 1974 and become a life member of the Indian Mathematical Society in 1981. However, I only started attending the meetings from 1986.

Halmos was active in both the AMS and MAA, but towards the end, he was heavily involved with the latter. **My conjecture is that Halmos must have been ticked off by the 'research brass' of the AMS.** It perhaps showed up three years ago when Halmos and his wife donated 3 million dollars to the MAA for a mathematical sciences conference center, but nothing to the AMS. Halmos, as a believer in divinely mathematics, has set a benchmark in philanthropy.

A couple of days ago, I told a friend that Halmos was a Beatle amongst mathematicians. The Beatles wrote and composed their songs, created and set the lyrics, played and sang them. They danced and smoked to their tunes. **Finite dimensional** Halmos was a captivating speaker, solid researcher, enthusiastic teacher, choosy mentor, popular author of books, research monographs, provocative writer of articles on math education, and sparkling media person. Above all, he had a large heart for his friends, whether human beings, animals, or institutions. **His legacy is secure and enduring**.

(Nov 06, 2006)

ON SURVEY OF A MATHEMATICAL COURSE

MAT 712 (**Survey of Mathematical Problems II**) has been intellectually fun as the semester winds down. Carryn Bellomo, scheduled to teach this course as a follow up to MAT 711, graciously agreed to let me teach it. My return to full time faculty position on leaving the associate deanship was sudden in summer.

However, my association with this sequence goes back to five years when *Teaching Mathematics* was approved as the fourth concentration in the master's program. The other three are: **Pure Math**, **Applied Math** and **Applied Statistics**. It was a culmination of 25 years of efforts! Graduate degree in math education has a place in mathematics.

The objective of this sequence is to have an integrative approach to mathematical topics and strengthen connections with high school mathematics. Whereas, a text book is a must for the US students, but finding one for this sequence turned out very difficult. One may think like a get-well card for every disease, there is a textbook for every curricular needs. Textbooks in USA are published by volume to meet the needs of the largest number of users, and not of the smallest. Stillwell's book on *Mathematics and its History* was the closest that I could settle on after browsing the Lied Library and various online searches.

It is important to establish a correspondence with students and choice of topics. Eight students represented a very diverse math background and aptitude for mathematics. There being a separate graduate course in the *History of Mathematics*, this course was to have minimal touch of history. Having no specific math prerequisites for MAT 711-712, it meant customizing the material dynamically all along.

A challenge in this course is to find a mathematical yarn and stretch it with as many problems and concepts as possible. Doing problems of undergraduate math courses alone is to undermine its graduate standing. I constantly reminded the students that after the course, you would be able to talk of some new problems, approaches, concepts, and terminology.

With open course description, neither typical math home work could be assigned, nor papers submitted. After two weeks into the classes, I suggested weekly projects on the material discussed. Its benchmarks were organized class material, library/online research on related topics, and attempt on mathematics problems. At times, handouts on problems were distributed. I remained open to suggestions from the students.

Looking back at the projects done over the last 12 weeks, this idea turned out effective. They also provide a better course record. For me, it was an eye opener to know how much math is there on the internet. Two special features of some projects are classic video, ***Mathematical Mystery Tour***, and news reports on ***Nobel Prizes and Mathematics***.

(Nov 20, 2006)

WILL IT BE ON THE TEST?

"This semester I had an unusual experience teaching the first course (MATH 427) on Differential Equations (DE). Almost every day while discussing a problem, one or two students would raise the questions about long time taken to do it in the class means it should not be on a test. Or, if it involves remembering a result(s), then how can it be on a test? As a consequence, some students sitting right in the front row would stop taking notes. It boils down to doing only the problems, if they could be on the test."

Being the last day of the session, I shared it with seven students in a graduate course having four full time high school teachers and three teaching assistants. I said such remarks did affect my enthusiasm and the general mood of the class. How can you compel them to do demanding problems? Home works carry only 10% of the grade. As a matter fact, only 4-5 students out of 23 have attempted at least 60-70% problems from the textbook. 50 % students are math edu/math majors!

Students who often raise these questions are, otherwise, mature students. They are just not used to being challenged. It is creative to choose the test problems, as any one taking 30 minutes are ruled out. In a course of DE, problems involving Reduction of the Order, Undetermined Coefficients, Variation of Parameters, Series and Laplace Transform are time consuming. You see structure of the problems only when you have done plenty of them and spent hours on a few. In the class, 80 % students are working!

Going back in time is my forte. Twenty two years ago, when I taught this course, and used the 3rd edition of Boyce DiPrima (now 8th ed). The number of the students was 14 whereas the UNLV student body was 11,000. On sharing my 22-year old test with the students, it was clear that the problems then were more solid. When my son was majoring in Math at UNLV (nearly 20 years ago), I had asked if he had spent 1-2 days on any problem. When he said No, I told him that during my undergraduate days in India, it was common.

In fact, for a group of nine problems, the instructor discussed one problem after our two weeks of frustration. **It opened my windows of understanding!**

My graduate students were not at all surprised. One teaching in a private school with only 300 students put it in perspective by describing it an everyday school scene. He said in a school campus where one's voice can be heard from one end to the other, the students call each other on cell phones the moment one is out of sight. **Finding a friend or finding a solution of a math problem has to be instantaneous.** AAha, that is all I said.

In 20 years, the attention span has shrunk, at large. How long the illegals or the first generation immigrants will do the menial jobs and routine intellectual projects in the US before the social fabric breaks down? I consider myself a good motivator and entertaining instructor, but I feel bedeviled. The problem is local to UNLV. Instructors are known to teach to the tests when ever they are common finals. But this mind set cuts deeper.

Satish C. Bhatnagar
(Dec 07, 2006)

JOY OF KNOWING THY STUDENTS

This **Reflection** is like a corollary to a theorem. The main result can be stated as: Tell me what you are passionate about in life, and I shall try to relate it with mathematics. The objective is to motivate the students for the course material. Four weeks ago, I asked my students to write a few lines on a sheet and turn it in.

After all, how do you distinguish an instructor with 40+ years in teaching with one who has been into it for a few years? **Any activity done with full consciousness for a long time becomes a window of life**. It is a paradigm shift. Some people are also known to repeat one year forty times! Novelty in life does not come easy.

Teaching remains refreshing with new crops of students irrespective of the course being the same. On the other hand, research stays stimulating with a movement of newer ideas and concepts. For me, knowing the students by their names and majors has been important. However, understanding their ardent hobbies or jobs goes a longer way.

Reading brief synopses of their engaging pursuits truly define UNLV students. First, these students are not 'youngish'. They come back to classrooms after a first hand experience of life. Here go my 21 students, mostly juniors and seniors, in a Differential Equations course (MATH 427). Rachel majors in geology, works in a geochemistry lab, and 'loves hiking as geologist'. Tim 'loves piano music, but not its practice'. Kelly is best as a basketball computer statistician. Amanda is 'passionate about Italian wines'!

Mohamed has enjoyed marital arts as student and instructor for 15 years. Jim coaches ice hockey. It is 'played with hands, legs, head and heart'. He focuses on 'head' and compares hockey with jazz! Corine loves finance. Robert, physics major, also digs into astronomy. Mary is a mother of two kids under 4. 'My greatest joy is being a mom'. She is amongst the top five. Pat, 20, youngest (oldest 40+) in the class, loves fishing. Shawn is 'really excited about making money and lot of it'. Diana played clarinet for 9 years, but truly enjoys math now! Carlee is 'passionate about soccer, but has no time with two jobs and fulltime student'. Lars loves sailing for its 'soothing effect', and it shows up.

David learns different languages. For Sara, dancing is 'like oxygen to the body; brings me to life'. Brian is avid about playing golf besides fulltime dealing in a casino. Adam can dis-assemble any electronic or mechanical device and put it back together, thus understands its working. He alone was perfect on the last quiz. Megan enjoys doing and tutoring math; is tied at the top. Dalia loves to count money, and once did count one million$$! June, mother of a H/S daughter, 'loves open water dives'. Her fitness shows up in her arms and flexing biceps. Ian who did not turn in, is perhaps still searching!

I enjoy teaching at UNLV. No where else in the world, you can encounter such a group of mathematics students. **The essence of teaching lies in enriching each other's life**!

(Dec 10, 2006)

Robert P. Gilbert And My PhD Saga
(On his 75th Birth Anniversary)

"The dissertation part of PhD degree in mathematics (a.e.) (Analysis notation for **almost everywhere** with sets of measure zero) is 25% perspiration and 75% inspiration, guidance, and the genial nature of the advisor. And I am so fortunate in having more than that in the person of Professor Robert P. Gilbert!" These are the first 4 lines of the 7-line *ACKNOWLEDGEMENTS* in my doctoral dissertation finally finished at Indiana University (IU) in Jan 74. I was Gilbert's last student at IU that he left in 1975.

My pursuit of doctorate is like present state of marriages! First is like one night stand type of relationship, second lasting for a couple of years. Finally, after some wising up, one seriously commits in the relationship. It was after two years of doing master's, in 1961 from Panjab University (PU), Chandigarh that the first thoughts of doing PhD came up. I was then in PU Evening College, Shimla. Most of the colleagues in social studies and languages were either planning or already registered for PhDs. It essentially requires that you approach a PhD professor in a university your college is affiliated with, and have a plan of research studies leading up to a thesis approved.

The PhD work mainly went on through correspondence and occasional visits. It is like modern online or distance education. There being no course work requirements, one finishes PhD in 3-4 years on the basis of thesis alone. It still goes on in most colleges and universities in India and neighboring countries, the erstwhile British colonies. **There is relatively little breadth, but more depth in this doctoral system.**

In the US, thesis is **"Submitted to the Faculty of the Graduate School in partial fulfillment of the requirements for the degree of doctor of Philosophy....."**. The other requirements are solid courses for 2-3 years, foreign language, comprehensive written exams, and finally, the oral defense of the thesis.

I had perhaps the highest score in number theory paper (two semester sequence) taught by eminent number theorist, R P Bambah. He used the classical textbook by Hardy and Wright. During a rigorous proof of Prime Number Theorem spanning over a week, once I pointed out a short cut in a section of its long proof. Bambah was surprised and remarked, "How come Hardy missed it!" It was during spring of 1961, and this moment stands out as one of the highlights of my studentship.

I was recommended to study Modern Algebra starting with P S Alexanderoff's book. Married a few months earlier, not much studying was done after the college. Also, a gap of two years is a long lay off to slacken the mathematical connections. I did MA at 21, met the most beautiful girl at 22, and married her after six months. Sex was far more possessive and exciting than Group Theory!

Still I was determined to get a PhD. During a winter break, I spent 2-3 weeks on PU campus to get over a hump in the material and gain some traction and momentum. One needs to do and discuss math in a right environment to understand the basics of the topics. Mathematics is very different from humanities and social studies. For that reason, I seldom encourage students to take regular courses under independent studies. It also dilutes the academic standards.

My wife has never forgotten and forgiven my 'missing' our first marriage anniversary falling during this period. Now I can state this reality like a mathematical theorem: Behind every man's PhD in mathematics, there is a woman; mother, wife, or girl friend. It is four years of deprivation and hard work. Framed PHT (Putting Husband Through) 'diplomas' are presented to wives when the husbands get their PhDs. **By way of inverse logic, neither have I heard a man behind a woman's PhD, nor seen a corresponding PWT diplomas for husbands!**

During summer of 1964, I met S D Chopra in Shimla. In PU, he taught a paper on *Electricity, Magnetism and Vectorial Calculus*. His colorful use of phrases during lectures has influenced my instructional style. He encouraged me for doing PhD in Mathematical Seismology at Kurukshetra University (KU) where he had joined as Professor and Head. Chopra brought Seismology to KU after his PhD from Cambridge University (UK) under Sir E R Lapwood, further a student of Sir Harold Jeffreys.

Leaving pregnant wife in Patiala with her parents, I immersed into the study of Continuum Mechanics and propagation of earthquake waves in layered media with different sources. It was April 1965 and I set a high goal of finishing PhD in 2-3 years! However, near the end of two years, suddenly my heart was not there. There were a few reasons; some friends went to USA and Canada for PhD, IU offered me an assistantship that I had applied for before joining KU, thesis not in sight, I felt a void in the knowledge of modern mathematics. With a new baby in the family, financial crunch bothered me on the top of emotional drain.

In 1967, a new college opened up in Patiala only 200 yards away from the house we rented. Essentially, I was one-man dept head with first two year of classes. I often said that if there is one place where I would settle in India, then it would be Patiala, a historic and beautiful city in Punjab! **Joyous living is not a function of one's will alone.** Within a few months, the clashes between the new faculty and the management erupted. By Spring 1968, the atmosphere became hellish. IU renewed the assistantship, and I left India in Aug 1968.

Some patterns of life are hard to explain. My wife was pregnant when I went to KU, and so was she again, when I left for IU! I stayed alone in KU, but every 2-3 weeks visited the family in Patiala and released sexual tension. For a year at IU, I was without family. It was hard to explain it to the Americans about some strength of Indian family system. Despite frugal living, I cleaned the apartments during summer months to save money for family airfares; though a loan for my airfare was still unpaid.

With family came the responsibly of two children and wife on the top of new education system and culture. The life style in India was that once you have done BA/MA, got a job and married, then that is end of studying. For me, it was quite a mental adjustment to be a student in the trenches again.

After many years of living in USA, I have come to realize that learning new ideas and skills is a life long activity. The constraints in India are due to its long colonial subjugation and poor economic conditions. Nevertheless, it defines two cultures and nations. I often think that a byproduct of doing PhD in mathematics from USA, varied experiences gained are enough to qualify me for a Ph D in 'humanities' too.

Before joining IU, I was familiar with the names of faculty, Tracy Thomas in Continuum Mechanics and Vaclav Halavaty (Einstein's student) in Relativity, and I desired to do PhD with one of them. But they had retired before I arrived. After a year of DE courses with E. Hopf (of Hopf-Weiner Equation and Hopf Maximal Principle) and written comprehensive exams, I approached Hopf for doing thesis under his guidance. Hopf was a very gentle and unassuming mathematician. He frankly told me that he was too old and had run out of research ideas. He spoke with Gilbert about me. As a matter of fact, two years earlier, Gilbert had bailed out Hopf's student, Ramankutty when in the middle, Hopf could not stimulate his thesis research.

I have known some young faculty coming out of Graduate schools and taking on PhD students that they cannot guide simply because they have not independently researched deep enough. During my tenure at IU, 3-4 students left it after 4-5 years without PhD! Yes, PhD is not like BA/MA degree that requires just the course work. To be able to think, identify new problems, and independently solve them, is a mark of PhD material. **During moments of research frustration, thesis supervisor and student both should not fail to overcome it.** That is where Gilbert stood out amongst the IU faculty.

Well, that is how I started working with Gilbert in 1971. Dean Kukral and Pat Brown were there along with a faculty member, David Colton in function theoretic PDE research group, as it was called. Gilbert having done his PhD with Zeev Nehari of complex analysis fame extended Bergman and Vekua's researches in several directions. Pat and I used to joke that Gilbert and Colton have emptied the entire vein in this gold mine! However, Gilbert had G. Hile work on hyperanalyitc function theory, a different research topic.

After a year, as I was getting ready to work on a thesis problem, Gilbert suddenly decided to go to Germany for a year. It was his getting away from a bitter marital separation. He frankly told me either to change the supervisor, or follow him to Germany, or stay at IU and communicate by mail. It was out of question to uproot the new family and tent it for a year in another culture. I decided to stay back and so did Pat as his wife, who was a doctoral student in English.

Despite initial gloominess, research-wise, 1972-73 turned out to be a banner year, as independently, I was able to prove some results using fundamental solutions of certain PDE. When Gilbert returned, I presented him a 'draft' of my work. He went over and told me to publish them later on. However, he wanted me to extend a few results on Pseudoparabloic PDE. I plunged into them during that summer.

The oil embargo, started in 1971, hit the US economy hard in 1973. I sent job applications to more than 200 non-PhD granting mathematics departments, but not even a single offer came from the US. I remained on student visa by delaying the thesis submission. Gilbert put me on his research grant.

During 1973-74, I also strengthened my credentials by taking courses in an emerging area of computer science. It paid off next year in 1974 as UNLV was looking for a person in applied mathematics and numerical analysis with computer science background. The Department had two failed searches. It was on the basis of my exceptional qualifications and Gilbert's strong letter of support that the INS labor certification was approved that eventually led up to my getting the US 'Green Card'.

Of all the non-mathematical anecdotes while working with him, a couple of them stand out. One evening, Gilbert got up to stretch, took out his American (oblong) football and we started playing throw and catch. The odds of hitting one of the four doors in a small hallway were high, and sure it went through one office glass door. **His 'genius' was to search, unhinge/hinge the broken door with a right size unbroken one!** It really impressed me.

Gilbert had a fancy German camera and drove a fancier BMW. I bought my first camera Konica S2 in 1969 and Ford Fairlane SW (1965) in 1971. Gilbert remains an avid photographer. Couple of years ago, I quit on taking other's pictures the moment I realized that **you are defined by the number of pictures others take of you!**

I took care of Gilbert's office when he was in Germany. In the process, I 'killed' his two ball cacti as I had no knowledge of the cactus care. Instead of watering them once a week, I did it every day! Well, it was a turning point in my cactus education. During 1980-82 stay in India, I became an expert on cacti and succulent plants, and had over 200 of them! Years ago, in Las Vegas, ours was the only front yard in the neighborhood with desert landscape and cacti that now are a rage.

Life is all about memories. **Immortality is an infinite sequence of memorable events.** Writings research papers are as good as they are remembered (referred). Teaching is good as long as the students remember you or recall your impact. Professionally, Gilbert has demonstrated his uncanny ability to do research with anyone around him. Nearly 300 research papers and 15 books and monographs testify to his individual intellect, research diversity and collaborative personality. Above all, he enjoys life 'beyond' the world of mathematics.

(Dec 16, 2006)